PARTIAL REGULARITY *for* HARMONIC MAPS *and* RELATED PROBLEMS

PARTIAL REGULARITY *for* HARMONIC MAPS *and* RELATED PROBLEMS

by

Roger Moser

Courant Institute of Mathematical Sciences

New York University, USA

World Scientific

NEW JERSEY • LONDON • SINGAPORE • BEIJING • SHANGHAI • HONG KONG • TAIPEI • CHENNAI

Published by

World Scientific Publishing Co. Pte. Ltd.

5 Toh Tuck Link, Singapore 596224

USA office: 27 Warren Street, Suite 401-402, Hackensack, NJ 07601

UK office: 57 Shelton Street, Covent Garden, London WC2H 9HE

Library of Congress Cataloging-in-Publication Data

Moser, Roger.

Partial regularity for harmonic maps and related problems/Roger Moser.

p. cm.

Includes bibliographical references and index.

ISBN 9812560858 (alk. paper)

1. Harmonic maps -- Mathematical models. 2. Mathematical physics. I. Title.

QA614.73.M67 2005

514'.74--dc22

2005041725

British Library Cataloguing-in-Publication Data

A catalogue record for this book is available from the British Library.

Printed in Singapore by World Scientific Printers (S) Pte Ltd

Preface

This monograph contains material from several research papers and lectures I gave in Bonn, Leipzig, New York, and Fribourg on various occasions, all of them about different aspects of the same problem. In an attempt to make the work nearly self-contained, I also included many additional paragraphs and most of the proofs of the auxiliary results. It is assumed, however, that the reader is familiar with the basic theory of linear elliptic and parabolic partial differential equations, and with the elementary notions of Riemannian geometry.

The aim of the book is to explain the methods that have been developed in the last decades to prove partial regularity for harmonic maps, and also to show how these methods can be extended to related problems. This includes perturbations of the harmonic map problem as well as associated parabolic problems. Both types may be of interest in applications from physics or possibly other sciences.

Roger Moser
New York
November 2004

Contents

Chapter 1

Introduction

Variational principles play an important role in both geometry and physics, and one of the key problems with applications in both fields is the variational problem associated to the Dirichlet energy of maps between Riemannian manifolds. The critical points of this functional are called harmonic maps.

Harmonic maps have successfully been used in many instances in order to understand the geometry of the involved manifolds. The Dirichlet energy also appears in the context of various theories from physics; for instance, in connection with liquid crystals, ferromagnetic materials, or superconductors. Another motivation to study harmonic maps is the fact that many principles that hold for the Dirichlet energy and their critical points, the harmonic maps, have counterparts in other theories. Methods that have been developed for harmonic maps can sometimes also be applied to other problems, and vice versa. For example, the problems of minimal surfaces or Yang-Mills fields show a similar behavior as the harmonic map problem in certain aspects.

The Dirichlet energy is one of the simplest possible functionals involving first derivatives. In the theory of harmonic maps, however, it is studied on a non-linear space, which gives rise to non-linear Euler-Lagrange equations for its critical points. These non-linearities make it a challenge to study the solutions. Whereas it is not difficult to construct weak solutions in general—for instance by the direct method from the calculus of variations—, in order to see how regular the weak solutions are, one needs quite subtle arguments. The methods that are used here take advantage of the special structure of the equations, and involve some rather deep results from harmonic analysis. It is the aim of this monograph to explain these arguments and to show how they can be used to prove regularity, or partial regularity (for the solutions are not regular everywhere in general) of solutions of related problems, including perturbations and generalizations of the harmonic map problem as well as parabolic problems associated to the Dirichlet energy.

1.1 Harmonic Maps

Suppose M and N are smooth Riemannian manifolds of dimensions m and k, respectively. We assume throughout the book that N is compact and has no boundary. The manifold M, in contrast, may be non-compact and may have a boundary. We consider maps $u : M \to N$, therefore we call M the domain and N the target manifold. We now give a definition of the Dirichlet energy of a sufficiently smooth map $u : M \to N$ (for instance, $u \in C^1(M; N)$). Because regularity is a local problem, and because the geometry of M is not so important for the arguments, we will assume later that M is an open subset of the Euclidean space $\mathbb{R}^m$ (and we will then write Ω instead of M). For completeness, however, we at least write down the problem for a general manifold M here.

A map $u : M \to N$ induces a metric on the vector bundle $T^*M \otimes u^{-1}TN$ over M (with fiber $T^*_xM \otimes T_{u(x)}N$ at the point $x \in M$). We write $\langle \cdot , \cdot \rangle_{T^*M\otimes u^{-1}TN}$ for this metric. If u is continuously differentiable on M, then the first derivative du is a continuous section of this vector bundle. Hence the function

$$e(u) = \frac{1}{2}|du|^2_{T^*M\otimes u^{-1}TN} = \frac{1}{2}\langle du, du\rangle_{T^*M\otimes u^{-1}TN}$$

is well-defined. We set

$$E(u) = \int_M e(u)\, d\mu_M, \tag{1.1}$$

where μ_M is the measure induced by the Riemannian metric on M, provided that the integral converges. This number is called the Dirichlet energy of the mapping u.

Next we give this expression in local coordinates. Suppose we work in a coordinate chart M' of M, and the metric is given there by the matrix $\gamma = (\gamma_{\alpha\beta})_{1\le\alpha,\beta\le m}$ in local coordinates. Suppose further that the image of M' under u is contained in a coordinate chart of N. The metric of N is given by $g = (g_{ij})_{1\le i,j\le k}$ in the corresponding local coordinates. We write $(\gamma^{\alpha\beta}) = \gamma^{-1}$, thus the metric of T^*M is given by this matrix. Then we have

$$e(u) = \frac{1}{2} \sum_{\alpha,\beta=1}^{m} \sum_{i,j=1}^{k} \gamma^{\alpha\beta} g_{ij} \frac{\partial u^i}{\partial x^\alpha} \frac{\partial u^j}{\partial x^\beta},$$

and the Dirichlet integral is given locally by the integral

$$E(u; M') = \frac{1}{2} \sum_{\alpha,\beta=1}^{m} \sum_{i,j=1}^{k} \int_{M'} \gamma^{\alpha\beta} g_{ij} \frac{\partial u^i}{\partial x^\alpha} \frac{\partial u^j}{\partial x^\beta} \sqrt{|\det \gamma|}\, dx.$$

When we compute the Euler-Lagrange equations for critical points of this functional, we find

$$\sum_{\alpha,\beta=1}^{m} \gamma^{\alpha\beta} \left(\frac{\partial^2 u^i}{\partial x^\alpha \partial x^\beta} - \sum_{\delta=1}^{m} \Gamma^{\delta}_{\alpha\beta} \frac{\partial u^i}{\partial x^\delta} + \sum_{j,l=1}^{k} C^i_{jl} \frac{\partial u^j}{\partial x^\alpha} \frac{\partial u^l}{\partial x^\beta} \right) = 0$$

for $i = 1, \ldots, k$. Here $\Gamma^{\delta}_{\alpha\beta}$ are the Christoffel symbols in M', and C^i_{jl} the Christoffel symbols for the local coordinates on N. If D denotes the Levi-Civita connection on $T^*M \otimes u^{-1}TN$, a coordinate free representation of these equations is

$$\mathrm{trace}(Ddu) = 0.$$

The point of view we have taken so far is not optimal when we consider the variational problem associated to the Dirichlet energy and weak solutions of the equations. Normally, it is natural to study such a functional in a Sobolev space; in this particular case, a space of maps with square integrable first derivatives would be an obvious choice. Unfortunately, maps in such a Sobolev space are not continuous in general, and the assumption that the image of M' under u is in a coordinate chart of N is not justified, even if M' is very small. In other words, we cannot carry out the same calculations as above for maps in a Sobolev space.

The easiest way to overcome this difficulty is to assume that N is isometrically embedded in a Euclidean space $\mathbb{R}^n$. This appears to be a restriction at a first glance, but in fact such an isometric embedding always exists, owing to the so-called Nash-Moser embedding theorem which was proved by J. Nash [1954; 1956] (and the methods were improved by J. Moser [1961]). We regard N as a submanifold of $\mathbb{R}^n$ in the rest of this book. We can then consider the Sobolev space

$$H^1(M; N) = \left\{ u \in H^1(M; \mathbb{R}^n) : \ u(x) \in N \text{ for almost every } x \in M \right\},$$

where $H^1(M;\mathbb{R}^n)$ is the Sobolev space of all weakly differentiable maps $u : M \to \mathbb{R}^n$ with

$$E(u) = \frac{1}{2}\int_M |du|^2_{T^*M\otimes\mathbb{R}^n}\, d\mu_M < \infty. \tag{1.2}$$

If u is in the space $H^1(M;N)$, then $du(x)$ is in $T^*_xM \otimes T_{u(x)}N$ at almost every point $x \in M$. Hence the number $E(u)$ defined by (1.2) coincides with the Dirichlet energy given by (1.1) whenever the latter is well-defined.

The new representation of the energy functional gives rise to a new version of the Euler-Lagrange equations. Suppose the second fundamental form of the submanifold N is denoted by A. Then it can be shown that critical points of E satisfy

$$\Delta^M u + \sum_{\alpha,\beta=1}^{m} \gamma^{\alpha\beta} A(u)\left(\frac{\partial u}{\partial x^\alpha}, \frac{\partial u}{\partial x^\beta}\right) = 0 \quad \text{on } M,$$

where $\Delta^M = \operatorname{div}^M(\nabla^M u)$ is the Laplace-Beltrami operator on M (and ∇^M denotes the gradient on M).

We assume henceforth that Ω is an open domain in $\mathbb{R}^m$, and we replace M by Ω. This simplification will spare us some technical work while preserving all the important ideas behind the arguments that are presented in the subsequent chapters. We then also use a simplified notation. The gradient on Ω is denoted by ∇, the Laplacian by Δ. We consider the Dirichlet energy

$$E(u) = \frac{1}{2}\int_\Omega |\nabla u|^2\, dx,$$

where $|\cdot|$ denotes the Euclidean norm in $\mathbb{R}^{m\times n} = \mathbb{R}^m \otimes \mathbb{R}^n$. The Euler-Lagrange equation for critical points of E becomes

$$\Delta u + A(u)(\nabla u, \nabla u) = 0 \quad \text{in } \Omega, \tag{1.3}$$

where we use the abbreviation

$$A(u)(\nabla u, \nabla u) = \sum_{\alpha=1}^{m} A(u)\left(\frac{\partial u}{\partial x^\alpha}, \frac{\partial u}{\partial x^\alpha}\right).$$

We will show in Section 3.2 how this equation is derived. (For a general domain manifold M, the computations are essentially the same.) It will also be shown that (1.3) is equivalent to

$$\Delta u(x) \perp T_{u(x)}N$$

for almost every $x \in \Omega$.

A map $u \in H^1(\Omega; N)$ which satisfies (1.3) in the weak sense is called a weakly harmonic map. The regularity of weakly harmonic maps is the first problem we study in this book. It turns out that weakly harmonic maps may have singularities in general. For instance, the map

$$u(x) = \frac{x}{|x|},$$

mapping $\mathbb{R}^m \backslash \{0\}$ onto the unit sphere

$$\mathbb{S}^{m-1} = \{x \in \mathbb{R}^m : \ |x| = 1\},$$

satisfies the equation

$$\Delta u + |\nabla u|^2 u = 0 \quad \text{in } \mathbb{R}^m \backslash \{0\}. \tag{1.4}$$

This, however, is (1.3) in the case $N = \mathbb{S}^{m-1}$ (as will be shown in Section 3.5). If $m \geq 3$, the map u satisfies (1.3) in the weak sense in $\mathbb{R}^m$. If it is restricted to a bounded domain $\Omega \subset \mathbb{R}^m$, it has finite energy and is thus a weakly harmonic map. But obviously u has a singularity at 0. The situation is even worse: There exists an example, constructed by T. Rivière [1995], of a weakly harmonic map that is discontinuous everywhere in its domain.

Fortunately, this is not the end of the story. On the one hand, the dimension of the domain plays a role here. For $m = 1$, we have an ordinary differential equation (and its solutions are geodesics). Regularity is not an issue here, and we don't even consider this case any more henceforth. The case of two-dimensional domains is special, too. We will see that for $m = 2$, every weakly harmonic map is smooth. The example of T. Rivière is in three dimensions, and from it, non-regular harmonic maps in higher dimensions can also be constructed. But despite these discouraging counter-examples, all is not lost even if $m \geq 3$. When we go back to the variational problem for the Dirichlet energy E, we find that it is natural to impose an additional condition, and then certain partial regularity results can be proved. Equation (1.3) is equivalent to the condition

$$\left.\frac{d}{ds}\right|_{s=0} E(\pi_N \circ (u + s\phi)) = 0$$

for any smooth map $\phi : \Omega \to \mathbb{R}^n$ with compact support, where π_N is the so-called nearest point projection onto N, i. e., the unique map that assigns to every point y in a neighborhood of N the point on N that minimizes the distance to y. The family of maps $\pi_N \circ (u + s\phi)$ is a natural variation of u in $H^1(\Omega; N)$, but this is not the only kind of variation we can consider. If we have a smooth map $\psi : \Omega \to \mathbb{R}^m$ with compact support, then the family given by $u_s(x) = u(x + s\psi(x))$ gives another variation of u. A weakly harmonic map $u \in H^1(\Omega; N)$ is called stationary if

$$\left.\frac{d}{ds}\right|_{s=0} E(u_s) = 0 \tag{1.5}$$

for all such variations. A smooth harmonic map is automatically stationary, but a weakly harmonic map not necessarily.

Stationary weakly harmonic maps can still have singularities. In fact the maps $u(x) = x/|x|$ discussed above are stationary for each $m \geq 3$. But in contrast to weakly harmonic maps in general, the singular set, i. e., the minimal relatively closed set in Ω such that u is smooth on the complement, is always small if the maps are stationary. For $u(x) = x/|x|$, the singular set consists of one single point. In general one can prove that a stationary weakly harmonic map is smooth away from a singular set of vanishing $(m-2)$-dimensional Hausdorff measure.

What makes stationary weakly harmonic maps special? When we want to prove regularity, the crucial observation is that the decay of the Dirichlet energy, restricted to concentric balls of shrinking radii, is described by a very convenient formula if we have a stationary weakly harmonic map. As a consequence, we have a good estimate for the mean oscillation (see the definition in Section 2.3) of the map. The problem of proving partial regularity of the kind as described above is then reduced to showing that a weakly harmonic map of small mean oscillation is always smooth. To do that, one can work with equation (1.3) again.

We now try to give a glimpse of how (1.3) can be exploited, with arguments based on ideas of F. Hélein [1990; 1991a; 1991b], in order to prove regularity. These arguments and the ones mentioned previously are given in detail in Chapter 3; here we just try to give an intuitive idea of how the structure of the equation comes into play. We consider again the special case where N is a sphere $\mathbb{S}^{n-1} \subset \mathbb{R}^n$. Then equation (1.3) can be written in the form (1.4), but there exists yet another equivalent system of equations: A map $u = (u^1, \ldots, u^n) \in H^1(\Omega; \mathbb{S}^{n-1})$ satisfies (1.4) if and only if

$$\operatorname{div}(u^i \nabla u^j - u^j \nabla u^i) = 0 \quad \text{in } \Omega \tag{1.6}$$

for all $i, j = 1, \ldots, n$. Using the exterior product $\wedge : \mathbb{R}^n \times \mathbb{R}^n \to \Lambda_2\mathbb{R}^n$, we can also write

$$\operatorname{div}(u \wedge \nabla u) = 0 \quad \text{in } \Omega$$

for (1.6). If we already know that u is regular, the equivalence of (1.4) and (1.6) can be checked directly, using the properties of the exterior product. It is shown in Section 3.5 that they are equivalent whenever $u \in H^1(\Omega; \mathbb{S}^{n-1})$. We have

$$|\nabla u|^2 = |u \wedge \nabla u|^2 = \sum_{i,j=1}^{n} u^i \nabla u^j \cdot (u^i \nabla u^j - u^j \nabla u^i),$$

because $u \perp \frac{\partial u}{\partial x^\alpha}$ almost everywhere for every $\alpha = 1, \ldots, m$. Thus the energy density

$$e(u) = \frac{1}{2}|\nabla u|^2$$

can be written in a very special form: the sum of products of

- a gradient,
- a divergence free vector field, and
- a function with a mean oscillation for which one has good estimates.

Then we need a few sophisticated tool from harmonic analysis. We will discuss them in the next section, along with some other analytic preliminaries. Here we just describe them briefly: A compensated compactness principle due to R. Coifman, P. L. Lions, Y. Meyer, and S. Semmes [1993] gives an estimate of the product of a gradient and a divergence free vector field in a Hardy space, in terms of the L^2-norms of the factors. This Hardy space is the dual to the space of functions of bounded mean oscillation, according to a result by C. Fefferman and E. M. Stein [1972]. In particular the product above has an estimate that allows to draw the following conclusion: If the energy of u is sufficiently small in a given ball, then the decay of the energy in concentric balls of shrinking radii is even better than the decay that condition (1.5) implies. In fact it is good enough to apply a well-known decay lemma of C. B. Morrey and conclude that u is Hölder continuous at least near the center of the ball. Continuity is the first step on the way to prove regularity, and it turns out that it is the most difficult. Once we know u is continuous, higher regularity is proved relatively easily.

For other target manifolds, this method does no longer work in the same way, because there is no representation of the harmonic map equation like

(1.6) in general. But with the help of additional arguments, due essentially to F. Hélein [1991a], similar arguments can still be applied.

1.2 Related Elliptic Equations

In applications, the harmonic map equation is often encountered not in its "pure" form, but with additional terms. For instance, the functional we study may be the Dirichlet energy plus some other, lower order terms. We consider briefly one such example as a motivation for the problems discussed in Chapter 4, namely an energy functional from the theory of micromagnetics. Many other functionals from physics or geometry could be considered instead, but we use this one because it is also relevant for another problem we will study later.

We assume $m = 3$. The bounded domain Ω represents the shape of a ferromagnetic sample and $u : \Omega \to \mathbb{R}^3$ its magnetization vector field. At low temperatures, one can assume that the magnitude of u is constant; for an appropriate choice of units this means $|u| = 1$ almost everywhere. That is, we have the target manifold $N = \mathbb{S}^2$. The energy assigned to u consists of several components:

- The highest order contribution is the Dirichlet energy, multiplied by a material constant d^2. That is, we have the term

$$\frac{d^2}{2} \int_\Omega |\nabla u|^2 \, dx.$$

  In this context, this is also called the exchange energy. The constant d is called the exchange length. The exchange energy penalizes spatial variations of the magnetization vector field, which reflects the fact that neighboring magnetization vectors have a tendency for parallel alignment.
- Crystalline anisotropies in the material can be modeled by a term of the form

$$\int_\Omega \Phi(u) \, dx$$

  for a function $\Phi : \mathbb{S}^2 \to \mathbb{R}$. We assume Φ is smooth.
- The magnetization induces a magnetic stray field $h : \mathbb{R}^3 \to \mathbb{R}^3$ which has a certain energy. According to the static Maxwell equa-

tions, the stray field satisfies

$$\begin{aligned} \operatorname{curl} h &= 0 \quad \text{in } \mathbb{R}^3, \\ \operatorname{div}(u+h) &= 0 \quad \text{in } \mathbb{R}^3, \end{aligned}$$

where u is extended by 0 outside of Ω. The first equation implies that $-h$ can be written as the gradient of a function U. By the second equation, this U is a solution of

$$\Delta U = \operatorname{div} u \quad \text{in } \mathbb{R}^3$$

in the distribution sense. There exists exactly one solution such that the integral

$$\frac{1}{2}\int_{\mathbb{R}^3} |\nabla U|^2\,dx = \frac{1}{2}\int_\Omega u\cdot\nabla U\,dx$$

is finite, and for this choice of U this integral gives the third contribution to the micromagnetic energy. It is called the magnetostatic energy.

- If there is an external field $H : \mathbb{R}^3 \to \mathbb{R}^3$, we also have the energy

$$-\int_\Omega u\cdot H\,dx.$$

The full micromagnetic energy is the sum of all four terms, that is, the functional

$$F(u) = \int_\Omega \left(\frac{d^2}{2}|\nabla u|^2 + \Phi(u) + \frac{1}{2}u\cdot\nabla U - u\cdot H\right)dx.$$

A discussion of the physical background of this energy can be found, e. g., in the books by A. Aharoni [2001] or by A. Hubert and R. Schäfer [1998].

Critical points of the functional F satisfy

$$d^2(\Delta u + |\nabla u|^2 u) - \operatorname{grad}\Phi(u) + (H-\nabla U)^\top = 0 \quad \text{in } \Omega,$$

where grad denotes the gradient on $\mathbb{S}^2$ and $(\,\cdot\,)^\top$ denotes the orthogonal projection $X^\top = X - \langle u, X\rangle\, u$ onto the tangent space of $\mathbb{S}^2$ at the image point of u. The first two terms are (up to the factor d^2) the same as in (1.4). They are the highest order contribution, hence they should be the most important terms when we study the regularity of solutions. It turns out that this is true. But when we try to prove regularity with the methods for harmonic maps, we have to handle quite subtle tools and difficult arguments, so that the additional terms may pose some problems, even if they are of lower order. Especially the term involving ∇U could cause some difficulties, because its non-local nature makes it difficult to say much about it in advance. Thus if we want to study the question of regularity for problems like this one, we need to find out how the methods can be adapted when we have variants of the harmonic maps equation.

In Chapter 4, we study solutions of the equation

$$\Delta u + A(u)(\nabla u, \nabla u) = f \quad \text{in } \Omega \tag{1.7}$$

for a function $f : \Omega \to \mathbb{R}^n$ in a certain L^p-space. In the example of the functional F, it is not difficult to show that the lower order terms of the equation are in appropriate L^p-spaces under reasonable assumptions, and the same is true for many other problems that can be encountered in applications.

It is sometimes relatively easy to generalize the methods of the theory of harmonic maps to such a situation, and sometimes very hard, depending mainly on the number p (in relation to the dimension m). We consider in particular three cases. First we study the case $p > \frac{m}{2}$ (and for technical reasons we always assume $p \geq 2$). Here we can prove Hölder continuity of solutions of (1.7) under the conditions that generalize the notion of stationary weakly harmonic maps naturally. In the border case $p = \frac{m}{2}$, continuity cannot be expected any longer; already the theory of linear equations shows that. It is still possible, however, to prove certain energy inequalities that can be useful for certain problems. Finally we study solutions of (1.7) for $p = 2$ in dimensions $m \leq 4$. If $p = 2$, the dimension four plays a similar role for the problem belonging to (1.7) as the dimension two plays for weakly harmonic maps. In particular, the problem is relatively easy—although obviously still more difficult than the corresponding problem for harmonic maps—if $m < 4$ (and hence $p > \frac{m}{2}$); it is moderately difficult if $m = 4$, and very difficult indeed if $m > 4$. We do not consider the last of these cases. In the other two cases, however, we can find some estimates for the second derivatives of solutions of (1.7).

There are many possible ways to generalize the arguments further. For instance, some of the methods we use in Chapter 4 also work for solutions of equations like

$$\Delta u + A(u)(\nabla u, \nabla u) = f + \operatorname{div} g \quad \text{in } \Omega,$$

where $g : \Omega \to \mathbb{R}^{m \times n}$ is in an appropriate L^q-space (and such equations may also appear in applications). Our goal, however, is not to give the most general results we can find, but rather to explain the methods. We could never cover all possible applications anyway.

When the results are finally applied to a problem like the one from micromagnetics discussed above, higher regularity of solutions can also be proved (under reasonable assumptions on Φ and H) away from the singular set. The analysis depends on the problem in question, however, and will therefore not be carried out in this book. Nevertheless it should be kept in mind that we are dealing here with what is in general the most difficult step towards proving regularity.

1.3 Evolution Problems

Methods to prove regularity for elliptic variational problems can often be modified so that they can be applied to the parabolic counterparts of the problems. In the case of harmonic maps and the related parabolic evolution problems, this strategy has its limitations, but in certain cases it is successful. We consider two types of evolution problems in Chapter 5: the so-called harmonic map heat flow and the Landau-Lifshitz (or Landau-Lifshitz-Gilbert) equation from the theory of ferromagnetism. The first can be seen as a general model for dynamic processes in physical systems of a certain type, but it has also been used in order to study harmonic maps and purely geometric questions. The Landau-Lifshitz equation can be regarded as a generalization of the harmonic map heat flow (for a special target manifold). The main motivation for its study, however, is its physical relevance. It also provides another example of how the methods developed for the theory of harmonic maps can give results for a wide variety of problems.

The harmonic map heat flow is the parabolic flow associated to the equation

$$\frac{\partial u}{\partial t} - \Delta u = A(u)(\nabla u, \nabla u) \quad \text{in } (0,T) \times \Omega \tag{1.8}$$

for some $T > 0$ (finite or infinite), where $u : (0,T) \times \Omega \to N$. For a more general domain manifold M, the corresponding equation is

$$\frac{\partial u}{\partial t} - \Delta^M u = \sum_{\alpha,\beta=1}^{m} \gamma^{\alpha\beta} A(u) \left(\frac{\partial u}{\partial x^\alpha}, \frac{\partial u}{\partial x^\beta} \right) \quad \text{in } (0,T) \times M'$$

in local coordinates on M', or

$$\frac{\partial u}{\partial t} - \operatorname{trace}(Ddu) = 0 \quad \text{in } (0,T) \times M$$

in a coordinate free form. We only consider the case $M = \Omega$ and the corresponding equation (1.8) for the same reasons as for the elliptic problems. The expression

$$\tau(u) = \Delta u + A(u)(\nabla u, \nabla u)$$

(called the tension field of u) is the negative L^2-gradient of the Dirichlet energy in the sense that

$$\left.\frac{d}{ds}\right|_{s=0} E(\pi_N \circ (u + s\phi)) = -\int_\Omega \langle \tau(u), \phi \rangle \, dx$$

for all $\phi \in C_0^\infty(\Omega; \mathbb{R}^n)$. In this sense the flow defined by (1.8) is the negative L^2-gradient flow of E. Accordingly one expects the energy inequality

$$E(u(t_2, \cdot)) + \int_{t_1}^{t_2} \int_\Omega \left| \frac{\partial u}{\partial t} \right|^2 dx \, dt \leq E(u(t_1, \cdot)) \tag{1.9}$$

for $0 < t_1 \leq t_2 < T$ under Dirichlet or Neumann boundary conditions. This is in fact true for smooth solutions, as the following arguments show: Taking the scalar product of both sides of (1.8) with $\frac{\partial u}{\partial t}$ yields

$$\left| \frac{\partial u}{\partial t} \right|^2 = \left\langle \Delta u, \frac{\partial u}{\partial t} \right\rangle,$$

because $\frac{\partial u}{\partial t} \perp A(u)(\nabla u, \nabla u)$. Hence

$$\frac{1}{2} \frac{\partial}{\partial t} |\nabla u|^2 = \left\langle \nabla u, \nabla \frac{\partial u}{\partial t} \right\rangle = \operatorname{div} \left\langle \nabla u, \frac{\partial u}{\partial t} \right\rangle - \left| \frac{\partial u}{\partial t} \right|^2.$$

Integrating this identity over Ω and integrating by parts, we find (1.9) in a differentiated form, if we have either Dirichlet boundary conditions and thus $\frac{\partial u}{\partial t} = 0$ on $(0,T) \times \partial\Omega$, or Neumann boundary conditions. It then only remains to integrate over the interval (t_1, t_2).

But for weak solutions in $H^1((0,T) \times \Omega; N)$, the energy inequality (1.9) is not true in general. The reason for this is related to the lack of regularity for weakly harmonic maps that are not stationary. Equation (1.8) alone does not imply that a weak solution behaves under variations of the domain the same way as a smooth solution does; but under an additional assumption—which we call the stability condition—that is the parabolic equivalent of

(1.5), the energy inequality holds, and so do other inequalities that are important when we want to prove regularity.

If the target manifold N is the sphere $\mathbb{S}^{n-1}$, the methods from the theory of harmonic maps extend quite naturally to the parabolic problem. We can prove that a weak solution of (1.8) satisfying the stability condition is smooth away from a set of vanishing m-dimensional parabolic Hausdorff measure (i. e., the Hausdorff measure with respect to the parabolic metric

$$d((t_1, x_1), (t_2, x_2)) = \max\{|x_1 - x_2|, \sqrt{|t_1 - t_2|}\}$$

on $(0,T)\times\mathbb{R}^m$). Since the time dimension behaves like two space dimensions in the parabolic metric, this means that the singular set is essentially at most of codimension 2, similarly as for harmonic maps.

The situation is different for general targets. Not that there is any reason to suspect that solutions of the harmonic map heat flow enjoy less regularity in this case; but for technical reasons, the arguments that work well for harmonic maps fail for the heat flow. The highly specialized tools we need, simply refuse to be bent into a different shape. For small dimensions of the domain, however, a slightly different approach is successful: Owing to a local version of the energy inequality (1.9) that we will derive in Section 5.1, we have

$$\frac{\partial u}{\partial t}(t, \cdot\,) \in L^2_{\mathrm{loc}}(\Omega; \mathbb{R}^n)$$

for almost every fixed $t \in (0, T)$. If we fix t, we can thus regard (1.8) as a version of (1.7). If $m \leq 4$, we have results from Chapter 4 that can be applied to this situation, and going back to space-time, we can then prove partial regularity for solutions of (1.8) under the stability condition.

The problem is open for $m \geq 5$. It appears that either new insights into the arguments or a completely different approach are needed to prove partial regularity for the harmonic map heat flow into general target manifolds in higher dimensions.

The other evolution problem we study, the Landau-Lifshitz equation, is a model for the dynamics of the magnetization vector field of a ferromagnet. It is related to the functional F introduced earlier. We will replace F by the Dirichlet energy E later, which simplifies the equation and makes the presentation of the methods we use clearer. For the moment, however, we give the full equation, in order to underscore its physical origin.

Suppose again $m = 3$. We write $\nabla_{L^2} F$ for the L^2-gradient of the functional F, i. e.,

$$\nabla_{L^2} F(u) = -d^2 \Delta u + D\Phi(u) + \nabla U - H.$$

Here we ignore the fact that F was defined only for maps into $\mathbb{S}^2$ previously; we extend Φ to $\mathbb{R}^3$ and write $D\Phi$ for its gradient in $\mathbb{R}^3$. (The Landau-Lifshitz equation, however, will only involve projections of this expression onto the tangent space of $\mathbb{S}^2$, therefore we could also use the L^2-gradient of the functional restricted to $H^1(\Omega; \mathbb{S}^2)$, instead.) We write $\wedge$ for the usual vector product in $\mathbb{R}^3$. Then the Landau-Lifshitz equation is

$$\frac{\partial u}{\partial t} = a\, u \wedge (u \wedge \nabla_{L^2} F(u)) + b\, u \wedge \nabla_{L^2} F(u) \quad \text{in } (0,T) \times \Omega. \tag{1.10}$$

Here $a > 0$ and $b \in \mathbb{R}$ are two fixed constants. The condition $a > 0$ makes the equation parabolic.

A solution u of (1.10) satisfies $u \perp \frac{\partial u}{\partial t}$, hence it follows that $\frac{\partial}{\partial t}|u|^2 = 0$. If we have initial data with values on the unit sphere $\mathbb{S}^2$, the solution remains on the sphere for all times. We only consider solutions with this property. Then (1.10) is equivalent to

$$a' \frac{\partial u}{\partial t} + b'\, u \wedge \frac{\partial u}{\partial t} = -(\nabla_{L^2} F(u))^\top \quad \text{in } (0,T) \times \Omega, \tag{1.11}$$

where

$$a' = \frac{a}{a^2 + b^2} \quad \text{and} \quad b' = \frac{b}{a^2 + b^2}.$$

Remember that $(\,\cdot\,)^\top$ denotes the orthogonal projection onto the tangent space of $\mathbb{S}^2$ at the point given by u. That is,

$$\begin{aligned}(\nabla_{L^2} F(u))^\top &= -d^2(\Delta u + |\nabla u|^2 u) \\ &\qquad + \operatorname{grad} \Phi(u) + \nabla U - H - \langle \nabla U - H, u \rangle\, u.\end{aligned}$$

It is explained in Section 5.5 how (1.11) is derived from (1.10) for the simplified equation (that is, for the functional E instead of F). The calculations for the full equation are essentially the same.

Now we replace F by the Dirichlet energy E. Then (1.11) becomes

$$a' \frac{\partial u}{\partial t} + b'\, u \wedge \frac{\partial u}{\partial t} = \Delta u + |\nabla u|^2 u \quad \text{in } (0,T) \times \Omega. \tag{1.12}$$

This equation makes sense in any dimension, not just $m = 3$. (But it should be kept in mind that the most relevant case in physics is $m = 3$.) The problem reminds very much of the harmonic map heat flow into $\mathbb{S}^2$;

indeed the heat flow is the special case of (1.12) corresponding to $a' = 1$ and $b' = 0$ (or equivalently, $a = 1$ and $b = 0$). For smooth solutions satisfying Dirichlet or Neumann boundary conditions, the energy inequality

$$E(u(t_2,\,\cdot\,)) + a' \int_{t_1}^{t_2} \int_\Omega \left| \frac{\partial u}{\partial t} \right|^2 dx\,dt \leq E(u(t_1,\,\cdot\,)) \tag{1.13}$$

holds for $0 < t_1 \leq t_2 < T$. This is shown with the same arguments as for the harmonic map heat flow, combined with the observation that

$$\left\langle a' \frac{\partial u}{\partial t} + b' u \wedge \frac{\partial u}{\partial t}, \frac{\partial u}{\partial t} \right\rangle = a' \left| \frac{\partial u}{\partial t} \right|^2 .$$

For weak solutions of (1.12), the energy inequality is no longer true, but there exists a generalization of the stability condition for the harmonic map heat flow which implies a local version of (1.13) and other estimates. In particular

$$\frac{\partial u}{\partial t}(t,\,\cdot\,) \in L^2_{\mathrm{loc}}(\Omega; \mathbb{R}^3)$$

for almost every fixed $t \in (0, T)$. If $m \leq 4$, we can apply the same methods as for the harmonic map heat flow and obtain the same kind of regularity result.

What about higher dimensions? The target manifold is $\mathbb{S}^2$ here, and we have no problems for the harmonic map heat flow into spheres, even in higher dimensions. Unfortunately, the flow belonging to the Landau-Lifshitz equation, being no gradient flow, behaves differently, and some of the arguments we use for the harmonic map heat flow do not work here. Again there is no reason to assume that the partial regularity results are false in higher dimensions, but a proof is lacking so far.

1.4 Notation

We now fix some notation that will be used frequently in the following chapters. Some of it has already been used above, but we repeat the definitions for completeness.

Throughout the book, Ω is an open subset of the Euclidean space $\mathbb{R}^m$, where $m \geq 2$. By N, we always denote a smooth, compact Riemannian manifold without boundary, which is isometrically embedded in the Euclidean space $\mathbb{R}^n$. We denote points of $\mathbb{R}^m$ in the form $x = (x^1, \ldots, x^m)$ and points of $\mathbb{R}^n$ by $y = (y^1, \ldots, y^n)$ or similarly. The Euclidean norm of a point x is denoted by $|x|$. We write $B_r(x_0)$ for the open ball in $\mathbb{R}^m$ of radius r around the point $x_0 \in \mathbb{R}^m$. If we have an open ball in another dimension, we write $B_r^j(x_0')$ instead, where j is the dimension (and x_0' is then in $\mathbb{R}^j$). The boundary of $B_r(x_0)$, that is, the sphere in $\mathbb{R}^m$ with radius r and center x_0, is denoted by $S_r(x_0)$. The unit sphere in $\mathbb{R}^n$ is denoted by $\mathbb{S}^{n-1}$.

An integral of the form

$$\int_\Omega f\,dx \quad \text{or} \quad \int_\Omega f(x)\,dx$$

is always meant with respect to the Lebesgue measure. The Lebesgue measure of a measurable set $\Omega' \subset \mathbb{R}^m$ is denoted by $|\Omega'|$. If S is a smooth $(m-1)$-dimensional hypersurface in $\mathbb{R}^m$, we indicate the area element on S by do. That is, an integral over S is written in the form

$$\int_S f\,do \quad \text{or} \quad \int_S f(x)\,do(x).$$

In a few cases, we use integrals with respect to a Hausdorff measure. We denote the j-dimensional Hausdorff measure by h^j; accordingly we write

$$\int_{\Omega'} f\,dh^j \quad \text{or} \quad \int_{\Omega'} f(x)\,dh^j(x)$$

for such an integral. We use two different notations for the average of a function f over a given measurable set Ω' (because each of them has its advantages), namely

$$\bar{f}_{\Omega'} \quad \text{and} \quad ⨍_{\Omega'} f\,dx.$$

That is, we write

$$\bar{f}_{\Omega'} = ⨍_{\Omega'} f\,dx = \frac{1}{|\Omega'|}\int_{\Omega'} f\,dx.$$

We use the abbreviation $\omega_j = h^j(B_1^j(0))$ for the j-dimensional volume of the j-dimensional unit ball; thus

$$\bar{f}_{B_r(x_0)} = \frac{1}{\omega_m r^m}\int_{B_r(x_0)} f\,dx.$$

We write $\nabla u = (\frac{\partial u}{\partial x^1}, \ldots, \frac{\partial u}{\partial x^m})$ for the (classical or weak) gradient of a function u on Ω, and $\nabla^k u$ for the k-th derivative.

By $\langle \cdot , \cdot \rangle$ we denote the standard scalar product in $\mathbb{R}^n$. Because N is isometrically embedded in $\mathbb{R}^n$, we can use these brackets also to indicate the scalar product of vectors in a tangential space T_yN, where $y \in N$. Sometimes we also use $\langle \cdot , \cdot \rangle$ to denote the scalar product in other Euclidean spaces. In $\mathbb{R}^m$, however, we indicate the standard scalar product by a dot. We use this notation to make it easier to understand expressions like

$$x \cdot \langle y, X \rangle \quad \text{or} \quad \langle y, x \cdot X \rangle ,$$

where $x = (x^1, \dots, x^m) \in \mathbb{R}^m$, $y = (y^1, \dots, y^n) \in \mathbb{R}^n$, and $X = (X_{\alpha i})_{1 \le \alpha \le m, 1 \le i \le n} \in \mathbb{R}^{m \times n}$. Here both expressions stand for the number

$$\sum_{\alpha=1}^{m} \sum_{i=1}^{n} x^\alpha y^i X_{\alpha i}.$$

When we study an evolution problem, we work in a set of the form $(0, T) \times \Omega$, where $0 < T \le \infty$. We denote points in $(0, T) \times \Omega$ by $z = (t, x) = (t, x^1, \dots, x^m)$. Instead of balls, it is usually convenient to consider parabolic cylinders of the form $P_r(z_0) = (t_0 - r^2, t_0 + r^2) \times B_r(x_0)$, where $z_0 = (t_0, x_0)$. Like in m dimensions, we write

$$\bar{f}_{P_r(z_0)} = \fint_{P_r(z_0)} f \, dz = \frac{1}{2\omega_m r^m} \int_{P_r(z_0)} f \, dz.$$

These parabolic cylinders are the equivalent of open balls for the parabolic metric

$$d((t_1, x_1), (t_2, x_2)) = \max\{|x_1 - x_2|, \sqrt{|t_1 - t_2|}\}.$$

We can also define a Hausdorff measure with respect to this metric in every dimension. We call this the parabolic Hausdorff measure. We will not need a symbol for the parabolic Hausdorff measure, but we will refer to it in some of the results in Chapter 5.

The symbol ∇ still denotes only derivatives with respect to the spatial coordinates $x^1, \dots, x^m$. The full derivative of a function u on $(0, T) \times \Omega$ is thus $(\frac{\partial u}{\partial t}, \nabla u)$.

We use the following classical function spaces:

- For $1 \le p < \infty$, $L^p(\Omega)$ is the space of all measurable functions u on Ω such that

$$\|u\|_{L^p(\Omega)} = \left(\int_\Omega |u|^p \, dx \right)^{1/p} < \infty.$$

The space $L^\infty(\Omega)$ consists of all essentially bounded measurable functions on Ω, and

$$\|u\|_{L^\infty(\Omega)} = \operatorname*{ess\,sup}_{\Omega} |u|.$$

- The Sobolev space $W^{k,p}(\Omega)$ is the space of all functions $u \in L^p(\Omega)$ with weak derivatives up to order k, such that

$$\|u\|_{W^{k,p}(\Omega)} = \sum_{l=1}^{k} \|\nabla^l u\|_{L^p(\Omega)} < \infty.$$

We write $H^k(\Omega) = W^{k,2}(\Omega)$.
- For $0 < \alpha \leq 1$, the Hölder space (or Lipschitz space if $\alpha = 1$) $C^{0,\alpha}(\overline{\Omega})$ consists of all continuous functions u on Ω such that

$$[u]_{C^{0,\alpha}(\overline{\Omega})} = \sup_{x,y\in\Omega,\ x\neq y} \frac{|u(x) - u(y)|}{|x-y|^\alpha} < \infty.$$

- The space $C^k(\Omega)$ consists of all continuous functions on Ω with continuous derivatives up to order k, and

$$C^\infty(\Omega) = \bigcap_{k=0}^{\infty} C^k(\Omega).$$

We write $C_0^\infty(\Omega)$ for the subspace consisting of all functions in $C^\infty(\Omega)$ with compact support in Ω.

The corresponding spaces for functions with values in $\mathbb{R}^n$ are denoted by $L^p(\Omega;\mathbb{R}^n)$, $W^{k,p}(\Omega;\mathbb{R}^n)$, etc. The subsets of these spaces consisting of all functions with values in N almost everywhere are $L^p(\Omega;N)$, $W^{k,p}(\Omega;N)$, etc. A similar notation is used for other target spaces. For the domain $(0,T)\times\Omega$, we have the spaces $L^p((0,T)\times\Omega)$, $W^{k,p}((0,T)\times\Omega)$, etc., defined similarly. The space $L^p_{\text{loc}}(\Omega)$ consist of all functions u such that for any compact set $\Omega' \subset \Omega$, we have $u \in L^p(\Omega')$; we use a similar notation for the local versions of the other function spaces. When one studies parabolic problems, it is natural to consider Hölder spaces that correspond to the parabolic metric d. We write

$$[u]_{C_d^{0,\alpha}([0,T]\times\overline{\Omega})} = \sup_{z_1,z_2\in(0,T)\times\Omega,\ z_1\neq z_2} \frac{|u(z_1) - u(z_2)|}{d(z_1,z_2)^\alpha}$$

and

$$C_d^{0,\alpha}([0,T] \times \overline{\Omega}) = \left\{ u \in C^0((0,T) \times \Omega) : \; [u]_{C_d^{0,\alpha}([0,T]\times\overline{\Omega})} < \infty \right\}.$$

We will also introduce and use other function spaces later on.

In order to simplify the proofs, we adopt the following convention: The symbol C is used to denote indiscriminately various constants that depend only on the dimension m of the domain Ω and the geometry of the target manifold N. Thus when we write an inequality of the form $a \leq Cb$ (for certain quantities a and b), we actually mean: "There exists a constant C, depending only on m and N, such that $a \leq Cb$ holds." Moreover, the same symbol C may denote different constants in the same proof or even in the same chain of inequalities. Constants that depend on other quantities as well are usually denoted by C_0, C_1, etc., or by a completely different symbol.

Chapter 2

Analytic Preliminaries

Before we examine the problems mentioned in the introduction in detail, we give a number of auxiliary results. Some are basic results from the theory of Sobolev spaces, like the Poincaré inequality or Morrey's decay lemma. We discuss them only briefly to remind the reader of the facts. We also present some deeper results in this chapter, however, for instance the BMO-Hardy space duality of C. Fefferman and E. M. Stein [1972]. Other things we discuss are a compensated compactness principle of R. Coifman, P. L. Lions, Y. Meyer, and S. Semmes [1993], a Hodge decomposition lemma due to T. Iwaniec [1992], a Gagliardo-Nirenberg type inequality of D. R. Adams and M. Frazier [1992], and a Gehring type lemma of M. Giaquinta and G. Modica [1979].

2.1 A Few Basic Inequalities Involving Sobolev Spaces

We assume that the reader is familiar with the Sobolev embedding theorem and the Poincaré inequality. It will be important, however, to know how the corresponding inequalities behave under a rescaling of the domain; therefore we say a few words about them anyway. By "rescaling", we mean a transformation of the form $x \mapsto rx + x_0$ for a fixed $x_0 \in \mathbb{R}^m$ and $r > 0$. If u is a function on Ω, this transforms u into the function $v(x) = u((x - x_0)/r)$ on $r\Omega + x_0$.

Now suppose Ω is the unit ball $B_1(0)$, and let $p \geq 1$. We have the Poincaré inequality

$$\int_{B_1(0)} |u - \bar{u}_{B_1(0)}|^p \, dx \leq C_1 \int_{B_1(0)} |\nabla u|^p \, dx$$

for a constant C_1 that depends only on m and p. For the rescaled function

v, this means

$$\int_{B_r(x_0)} |v - \bar{v}_{B_r(x_0)}|^p \, dx \leq C_1 r^p \int_{B_r(x_0)} |\nabla v|^p \, dx.$$

In fact these two inequalities are equivalent. To see this, we only have to check how the expressions transform under the rescaling. Similarly we obtain the Sobolev inequalities

$$\left(\fint_{B_r(x_0)} |v - \bar{v}_{B_r(x_0)}|^q \, dx \right)^{1/q} \leq C_2 \left(r^{p-m} \int_{B_r(x_0)} |\nabla v|^p \, dx \right)^{1/p}$$

for a constant C_2 that depends only on m and p, whenever $p < m$ and

$$1 \leq q \leq \frac{mp}{m-p}.$$

The Poincaré inequality can be used to prove another well-known result, namely the following decay lemma of C. B. Morrey [1966]. This lemma allows to derive Hölder continuity from an estimate of the Dirichlet energy (or similar expressions) on concentric balls.

Lemma 2.1 *Suppose $\alpha > 0$ and $p \geq 1$. There exists a constant C_0, depending only on m and α, such that the following holds. Suppose $u \in W^{1,p}(B_{2R}(x_0))$ satisfies*

$$r^{p-m-\alpha} \int_{B_r(x_1)} |\nabla u|^p \, dx \leq a \tag{2.1}$$

for every $x_1 \in B_R(x_0)$ and $0 < r \leq R$. Then for almost all $x_1, x_2 \in B_R(x_0)$,

$$|u(x_1) - u(x_2)| \leq C_0 a^{1/p} |x_1 - x_2|^{\alpha/p}.$$

Proof. First we note that the Hölder inequality implies

$$r^{1-m-\alpha/p} \int_{B_r(x_1)} |\nabla u| \, dx \leq \omega_m^{1-1/p} a^{1/p}$$

if (2.1) is true. Hence we may assume for the rest of the proof that $p = 1$.

The claim can now be reduced to the Poincaré inequality. For a given pair of points $x_1, x_2 \in B_R(x_0)$, set $r_0 = \frac{1}{2}|x_1 - x_2|$ and $\bar{x} = \frac{1}{2}(x_1 + x_2)$. Then we have

$$\left| \fint_{B_{r_0}(x_1)} u \, dx - \fint_{B_{r_0}(x_2)} u \, dx \right| \leq C r_0^{1-m} \int_{B_{2r_0}(\bar{x})} |\nabla u| \, dx \leq C a r_0^{\alpha}.$$

For $r_k = 2^{-k}r_0$, we have

$$\left| \fint_{B_{r_k}(x_1)} u\,dx - \fint_{B_{r_{k-1}}(x_1)} u\,dx \right| \leq C r_{k-1}^{1-m} \int_{B_{r_{k-1}}(x_1)} |\nabla u|\,dx \leq C a r_{k-1}^{\alpha},$$

again by the Poincaré inequality. The same estimates hold for x_2 instead of x_1. Since

$$\fint_{B_{r_k}(x_1)} u\,dx \to u(x_1) \quad \text{as } k \to \infty$$

by Lebesgue's differentiation theorem for almost every x_1 (and the same for almost every x_2), it only remains to sum up all these inequalities. We obtain

$$|u(x_1) - u(x_2)| \leq C a r_0^{\alpha} \left(1 + \sum_{k=0}^{\infty} 2^{-k\alpha} \right) \leq C_1 a r_0^{\alpha},$$

where C_1 is a constant that depends only on m and α. □

Remark 2.1 *Lemma 2.1 also gives a proof of a well-known Sobolev embedding theorem. Namely, the space $W^{1,p}(B_R(x_0))$ is embedded continuously in $C^{0,1-m/p}(\overline{B_R(x_0)})$ if $p > m$. Suppose $u \in W^{1,p}(B_R(x_0))$; then it can be extended to $B_{2R}(x_0)$ such that*

$$\|u\|_{W^{1,p}(B_{2R}(x_0))} \leq C_1 \|u\|_{W^{1,p}(B_R(x_0))}$$

for a constant C_1 that depends only on m and p. Estimate (2.1) is then satisfied for $\alpha = p - m$. It follows that

$$[u]_{C^{0,1-m/p}(\overline{B_R(x_0)})} \leq C_2 \|\nabla u\|_{L^p(B_R(x_0))}$$

for a constant C_2 that depends only on m and p.

Similar observations can be made for parabolic cylinders instead of balls. Here the right kind of rescaling is of the form $(t, x) \mapsto (r^2 t + t_0, rx + x_0)$. This maps the parabolic cylinder $P_1(0)$ onto $P_r(t_0, x_0)$, and a function u on $P_1(0)$ is transformed into $v(t, x) = u((t - t_0)/r^2, (x - x_0)/r)$ on $P_r(t_0, x_0)$.

The inequality

$$\int_{P_1(0)} |u - \bar{u}_{P_1(0)}|^p\,dz \leq C_1 \int_{P_1(0)} \left(|\nabla u|^p + \left| \frac{\partial u}{\partial t} \right|^p \right) dz,$$

for a constant C_1 that depends only on m and p, is just another variant of the Poincaré inequality in $m + 1$ dimensions. We can apply this inequality to the rescaled function v. Keeping track of the scaling factors, we find that

$$\int_{P_r(z_0)} |v - \bar{v}_{P_r(z_0)}|^p \, dz \leq C_1 r^p \int_{P_r(z_0)} \left(|\nabla v|^p + r^p \left| \frac{\partial v}{\partial t} \right|^p \right) dz.$$

Therefore we can prove the following parabolic version of Morrey's decay lemma with exactly the same method that we used in the proof of Lemma 2.1. We leave the details to the reader.

Lemma 2.2 *Suppose $\alpha > 0$ and $p \geq 1$. There exists a constant C_0, depending only on m and α, such that the following holds true. If $u \in W^{1,p}(P_{2R}(z_0))$ satisfies*

$$r^{p-m-2-\alpha} \int_{P_r(z_1)} \left(|\nabla u|^p + r^p \left| \frac{\partial u}{\partial t} \right|^p \right) dz \leq a$$

for every $z_1 \in P_R(z_0)$ and $0 < r \leq R$, then

$$|u(z_1) - u(z_2)| \leq C_0 a^{1/p} \left(|x_1 - x_2|^{\alpha/p} + |t_1 - t_2|^{\alpha/(2p)} \right)$$

for almost all $z_1 = (t_1, x_1), z_2 = (t_2, x_2) \in P_R(z_0)$.

Remark 2.2 *For $p > m$, we conclude that*

$$[u]_{C_d^{0,1-m/p}(\overline{P_R(z_0)})} \leq C_1 \left(\|\nabla u\|_{L^p(P_R(x_0))} + R \left\| \frac{\partial u}{\partial t} \right\|_{L^p(P_R(x_0))} \right)$$

for a constant C_1 that depends only on m and p.

2.2 The Hodge Decomposition

Consider a vector field $X : \mathbb{R}^m \to \mathbb{R}^m$. Suppose the L^p-norm of X is finite. We want to decompose such a vector field into the sum of a gradient vector field and a divergence free vector field. Moreover, the decomposition has to be continuous with respect to the L^p-norm. The existence of such a decomposition was proved by T. Iwaniec [1992] (see also the papers of T. Iwaniec and G. Martin [1993] and T. Iwaniec and C. Sbordone [1994]). It is called the Hodge decomposition.

Before we present these results, we introduce some more notation. For $1 < p < \infty$, the homogeneous Sobolev space $\dot{W}^{1,p}(\mathbb{R}^m)$ is the closure of

$C_0^\infty(\mathbb{R}^m)$ with respect to the topology given by the norm

$$\|u\|_{\dot{W}^{1,p}(\mathbb{R}^m)} = \|\nabla u\|_{L^p(\mathbb{R}^m)}.$$

If $\Omega \subset \mathbb{R}^m$ is an open set, then $W_0^{1,p}(\Omega)$ is the closure of $C_0^\infty(\Omega)$ with respect to the $W^{1,p}$-topology.

Theorem 2.1 *Let $1 < p < \infty$ and $X \in L^p(\mathbb{R}^m; \mathbb{R}^m)$. Then there exist a function $u \in \dot{W}^{1,p}(\mathbb{R}^m)$ and a vector field $Y \in L^p(\mathbb{R}^m; \mathbb{R}^m)$ satisfying $\operatorname{div} Y = 0$ in $\mathbb{R}^m$ in the distribution sense, such that*

$$X = \nabla u + Y.$$

There exists a constant C_0, depending only on p and m, such that

$$\|\nabla u\|_{L^p(\mathbb{R}^m)} + \|Y\|_{L^p(\mathbb{R}^m)} \leq C_0 \|X\|_{L^p(\mathbb{R}^m)}. \tag{2.2}$$

Proof. Let $u \in \dot{W}^{1,p}(\mathbb{R}^m)$ be the unique solution of the equation

$$\Delta u = \operatorname{div} X \quad \text{in } \mathbb{R}^m.$$

That is, if

$$\Gamma(x) = \begin{cases} -\dfrac{1}{2\pi} \log |x| & \text{if } m = 2, \\ \dfrac{|x|^{2-m}}{m(m-2)\omega_m} & \text{if } m \geq 3, \end{cases}$$

is the fundamental solution of Laplace's equation, we set

$$u(x) = \int_{\mathbb{R}^m} \nabla \Gamma(x-y) \cdot X(y)\, dy.$$

Then the vector field $Y = X - \nabla u$ clearly satisfies

$$\operatorname{div} Y = 0 \quad \text{in } \mathbb{R}^m.$$

Inequality (2.2) follows from standard L^p-estimates for singular integrals (see, e. g., E. M. Stein [1970]). □

We will also use a local version of the Hodge decomposition. We assume in the following that we have a vector field on a ball $B_r(x_0) \subset \mathbb{R}^m$. Similar results can also be proved for other domains, but we will not need them.

Theorem 2.2 *Let $1 < p < \infty$ and $B_r(x_0) \subset \mathbb{R}^m$. For every $X \in L^p(B_r(x_0); \mathbb{R}^m)$ there exist a function $u \in W_0^{1,p}(B_r(x_0))$ and a vector field*

$Y \in L^p(B_r(x_0); \mathbb{R}^m)$ satisfying $\operatorname{div} Y = 0$ *in $B_r(x_0)$ in the distribution sense, such that*

$$X = \nabla u + Y.$$

There exists a constant C_0, depending only on p and m, such that

$$\|\nabla u\|_{L^p(B_r(x_0))} + \|Y\|_{L^p(B_r(x_0))} \le C_0 \|X\|_{L^p(B_r(x_0))}. \tag{2.3}$$

Proof. The arguments are practically the same as for the previous theorem. Let $u \in W^{1,p}(B_r(x_0))$ be the unique solution of

$$\begin{cases} \Delta u = \operatorname{div} X & \text{in } B_r(x_0), \\ u = 0 & \text{on } \partial B_r(x_0). \end{cases}$$

Then the vector field $Y = X - \nabla u$ satisfies

$$\operatorname{div} Y = 0 \quad \text{in } B_r(x_0).$$

Estimate (2.3) follows immediately from standard L^p-estimates for elliptic partial differential equations (see, e. g., Chapter 9 of D. Gilbarg and N. S. Trudinger [1983]). □

2.3 Bounded Mean Oscillation

The notion of functions of bounded mean oscillation goes back to a paper of F. John and L. Nirenberg [1961]. The space of such functions can be thought of as an intermediate space which is a little larger than $L^\infty(\Omega)$, yet smaller than

$$\bigcap_{1 \le p < \infty} L^p_{\text{loc}}(\Omega).$$

Definition 2.1 Suppose u is a measurable function on Ω. If $u \in L^1(B_r(x_0))$ for every ball $B_r(x_0) \subset \Omega$, then we define

$$[u]_{\text{BMO}(\Omega)} = \sup_{B_r(x_0) \subset \Omega} \fint_{B_r(x_0)} |u - \bar{u}_{B_r(x_0)}| \, dx.$$

Otherwise we set $[u]_{\text{BMO}(\Omega)} = \infty$. The space of functions of *bounded mean oscillation* is

$$\text{BMO}(\Omega) = \left\{ u \in L^1_{\text{loc}}(\Omega) : \; [u]_{\text{BMO}(\mathbb{R}^m)} < \infty \right\}.$$

Remark 2.3 *Because it vanishes on constant functions,* $[\,\cdot\,]_{\mathrm{BMO}(\Omega)}$ *is only a seminorm, not a norm. The space* BMO(Ω) *can be made a Banach space, however, by taking the quotient with the subspace of constant functions.*

For vector-valued functions, the BMO-spaces and the corresponding seminorms are defined by the obvious generalization. We write BMO($\Omega; \mathbb{R}^n$), etc.

Sometimes it is convenient to work with cubes instead of balls. In the following, we write

$$Q_r(x_0) = (x_0^1 - r, x_0^1 + r) \times \ldots \times (x_0^m - r, x_0^m + r)$$

for $x_0 = (x_0^1, \ldots, x_0^m) \in \mathbb{R}^m$ and $r > 0$. Whenever we use the word "cube", we mean a cube of this kind, that is, one with sides parallel to the coordinate axes. Note that for any $u \in L^1(Q_r(x_0))$, we have

$$\begin{aligned}
\fint_{B_r(x_0)} |u - \bar{u}_{B_r(x_0)}|\,dx &\leq \frac{2^m}{\omega_m} \fint_{Q_r(x_0)} |u - \bar{u}_{B_r(x_0)}|\,dx \\
&\leq \frac{2^m}{\omega_m} \left(\fint_{Q_r(x_0)} |u - \bar{u}_{Q_r(x_0)}|\,dx \right. \\
&\qquad \left. + |\bar{u}_{B_r(x_0)} - \bar{u}_{Q_r(x_0)}| \right).
\end{aligned}$$

Moreover,

$$\begin{aligned}
|\bar{u}_{B_r(x_0)} - \bar{u}_{Q_r(x_0)}| &= \left| \fint_{B_r(x_0)} (u - \bar{u}_{Q_r(x_0)})\,dx \right| \\
&\leq \frac{2^m}{\omega_m} \fint_{Q_r(x_0)} |u - \bar{u}_{Q_r(x_0)}|\,dx.
\end{aligned}$$

Thus whenever we have a compact set $\Omega' \subset \Omega$, there exists a constant C_0, depending only on m, Ω, and Ω', such that

$$\sup_{B_r(x_0) \subset \Omega'} \fint_{B_r(x_0)} |u - \bar{u}_{B_r(x_0)}|\,dx \leq C_0 \sup_{Q_r(x_0) \subset \Omega} \fint_{Q_r(x_0)} |u - \bar{u}_{Q_r(x_0)}|\,dx.$$

Similarly, for the appropriate choice of C_0, we have

$$\sup_{Q_r(x_0) \subset \Omega'} \fint_{Q_r(x_0)} |u - \bar{u}_{Q_r(x_0)}|\,dx \leq C_0 \sup_{B_r(x_0) \subset \Omega} \fint_{B_r(x_0)} |u - \bar{u}_{B_r(x_0)}|\,dx.$$

The same holds for $\Omega = \Omega' = \mathbb{R}^m$. Hence for most purposes, we can either work with balls or with cubes, whichever is more convenient. We will take advantage of this fact later in this section.

First, however, we recall the classical Calderón-Zygmund cube decomposition lemma. We include a sketch of the proof for the convenience of the reader.

Lemma 2.3 *Suppose $u \in L^1(Q_1(0))$ is a non-negative function and $s > 0$ a number satisfying*

$$\fint_{Q_1(0)} u\,dx \leq s.$$

Then there exists a finite or countably infinite collection $\{Q_i\}$ (where i ranges between 1 and $I \leq \infty$) of pairwise disjoint cubes, such that

$$s < \fint_{Q_i} u\,dx \leq 2^m s$$

for each i, and $u \leq s$ almost everywhere in

$$Q_1(0) \setminus \bigcup_{i=1}^{I} Q_i.$$

Proof. We divide $Q_1(0)$ into 2^m cubes of side length 1 by a bisection of the edges. We include in our collection all cubes Q of this decomposition that satisfy

$$\fint_Q u\,dx > s. \tag{2.4}$$

All remaining cubes are divided further, into 2^m subcubes each, by the same method. We add to our collection all of the newly obtained cubes Q which satisfy (2.4). Again all other cubes are further subdivided, and this procedure is repeated indefinitely.

Thus $\{Q_i\}$ consists of all cubes Q obtained in one of the infinitely many steps that satisfy (2.4). By construction, they have all been obtained by subdivision of a cube that does not satisfy (2.4). This implies the first property we have to show. To prove the other one, we observe that for almost all points x in

$$Q_1(0) \setminus \bigcup_{i=1}^{I} Q_i,$$

there exists a sequence of cubes converging to $\{x\}$ that do not satisfy (2.4). Lebesgue's differentiation theorem implies $u(x) \leq s$ for almost all such points. □

We use this lemma in order to prove a property of functions of bounded mean oscillation that has first been found by F. John and L. Nirenberg [1961]. The arguments we give here are the same that have been used in that paper.

Lemma 2.4 *There exist two constants $b > 0$ and C, depending only on m, such that the following holds. Suppose $u \in L^1(Q_1(0))$ is a function such that*

$$\int_{Q_1(0)} u \, dx = 0$$

and

$$⨍_Q |u - \bar{u}_Q| \, dx \leq c$$

for all cubes $Q \subset Q_1(0)$. For $s > 0$, define

$$S(u, s) = \{x \in Q_1(0) : \ |u(x)| > s\} .$$

Then

$$|S(u, s)| \leq Ce^{-bs/c}$$

for all $s > 0$.

Proof. It suffices to prove the lemma for $c = 1$. For other values, the claim can be reduced to this special case by multiplying u by $1/c$.

Let $\mathcal{U}$ be the set of all functions $u \in L^1(Q_1(0))$ that satisfy the conditions of the lemma, that is,

$$\int_{Q_1(0)} u \, dx = 0$$

and

$$⨍_Q |u - \bar{u}_Q| \, dx \leq 1$$

for all cubes $Q \subset Q_1(0)$. Define

$$\alpha(s) = \sup_{u \in \mathcal{U}} |S(u, s)|$$

for $s > 0$. Obviously,

$$\alpha(s) \le \frac{2^m}{s}.$$

Suppose $u \in \mathcal{U}$. For any given number $s' \ge 1$, we can apply Lemma 2.3 to the function $|u|$ for s' instead of s. We obtain a collection $\{Q_i\}$ (for i between 1 and $I \le \infty$) of pairwise disjoint cubes such that

$$s' \le \fint_{Q_i} |u|\, dx \le 2^m s'$$

for each i, and $|u| \le s'$ almost everywhere in

$$Q_1(0) \backslash \bigcup_{i=1}^{I} Q_i.$$

If $s > 2^m s'$, it follows that

$$S(u,s) \subset \bigcup_{i=1}^{I} (Q_i \cap S(u,s)) \subset \bigcup_{i=1}^{I} (Q_i \cap S(u - \bar{u}_{Q_i}, s - 2^m s')).$$

Write $Q_i = Q_{r_i}(x_i)$. Then the functions

$$u_i(x) = u(r_i x + x_i) - \bar{u}_{Q_i}, \quad x \in Q_1(0),$$

belong to $\mathcal{U}$. Hence

$$|Q_i \cap S(u - \bar{u}_{Q_i}, s - 2^m s')| \le r_i^m \alpha(s - 2^m s'),$$

and we conclude that

$$\begin{aligned} |S(u,s)| &\le \alpha(s - 2^m s') \sum_{i=1}^{I} r_i^m \le \frac{\alpha(s - 2^m s')}{2^m s'} \sum_{i=1}^{I} \int_{Q_i} |u|\, dx \\ &\le \frac{\alpha(s - 2^m s')}{2^m s'} \int_{Q_1(0)} |u|\, dx \le \frac{\alpha(s - 2^m s')}{s'}. \end{aligned}$$

Hence

$$\alpha(s) \le \frac{\alpha(s - 2^m s')}{s'}$$

whenever $s' \ge 1$ and $s > 2^m s'$. In particular

$$\alpha(s + 2^m e) \le e^{-1} \alpha(s)$$

for every $s > 0$. Thus for $b = 2^{-m}e^{-1}$, the function α has the following property: if

$$\alpha(s) \leq Ce^{-bs}, \tag{2.5}$$

then

$$\alpha(s + 2^m e) \leq Ce^{-b(s+2^m e)}.$$

But (2.5) is obvious for $s \leq 1$, provided that $C \geq 2^m e^b$. The claim of the lemma thus follows by induction. □

Lemma 2.5 *For any compact set $\Omega' \subset \Omega$ and every $p \in [1, \infty)$ there exists a constant C_0, depending only on m, p, Ω, and Ω', such that any $u \in \mathrm{BMO}(\Omega)$ satisfies*

$$\|u - \bar{u}_{\Omega'}\|_{L^p(\Omega')} \leq C_0 [u]_{\mathrm{BMO}(\Omega)}.$$

Proof. It suffices to consider the case $\Omega = Q_2(0)$ and $\Omega' = \overline{Q_1(0)}$. In the more general situation the claim then follows by a covering argument. We may also assume that $\bar{u}_{Q_1(0)} = 0$ and $[u]_{\mathrm{BMO}(Q_1(0))} = 1$. (If $[u]_{\mathrm{BMO}(Q_1(0))} = 0$, then u is constant in $Q_1(0)$, and there is nothing to prove. Otherwise we can consider the function

$$\frac{u - \bar{u}_{Q_1(0)}}{[u]_{\mathrm{BMO}(Q_1(0))}}$$

instead of u.) If we can prove under these assumptions that

$$\|u\|_{L^p(Q_1(0))} \leq C_1$$

for a constant C_1 that depends only on m and p, then the proof is finished.

By Lemma 2.4, the sets

$$S(u, s) = \{x \in Q_1(0) : \ |u(x)| > s\}$$

satisfy

$$|S(u, s)| \leq Ce^{-bs}$$

for two constants C and $b > 0$ which depend only on m. Now we use the identity

$$\int_{Q_1(0)} |u|^p \, dx = \int_0^\infty |S(u^p, s)| \, ds = p \int_0^\infty s^{p-1} |S(u, s)| \, ds.$$

The first of these equalities follows from Fubini's theorem, the second from the substitution formula. We have

$$\int_0^1 s^{p-1}|S(u,s)|\,ds \le 2^m$$

and

$$\int_1^\infty s^{p-1}|S(u,s)|\,ds \le \int_1^\infty s^{p-1}e^{-bs}\,ds \le C_2$$

for a constant C_2 that depends only on m and p. Combining these inequalities, we obtain the desired estimate. □

Next we show that functions of bounded means oscillation behave well under multiplication with a C_0^∞-function. This will help to find localizations of certain inequalities involving the BMO-seminorm. The arguments we use here are due to L. C. Evans [1991].

Lemma 2.6 *Let $\eta \in C_0^\infty(\Omega)$ and $u \in \mathrm{BMO}(\Omega)$. Then the function $v = \eta(u - \bar{u}_\Omega)$ is in $\mathrm{BMO}(\mathbb{R}^m)$ with*

$$[v]_{\mathrm{BMO}(\mathbb{R}^m)} \le C_0[u]_{\mathrm{BMO}(\Omega)}$$

for a constant C_0 which depends only on Ω and η.

Proof. Choose an open set $\Omega' \subset \Omega$ such that $\overline{\Omega'} \subset \Omega$ and $\operatorname{supp}\eta \subset \Omega'$. By Lemma 2.5,

$$\|u - \bar{u}_{\Omega'}\|_{L^m(\Omega')} \le C_1[u]_{\mathrm{BMO}(\Omega)} \tag{2.6}$$

for a constant C_1 that depends only on m, Ω, and Ω' (and the choice of Ω' in turn depends only on Ω and η). Subtracting a constant from u changes neither $[u]_{\mathrm{BMO}(\Omega)}$ nor v, thus we can assume without loss of generality that $\bar{u}_{\Omega'} = 0$.

Suppose now $B_r(x_0) \subset \Omega'$. Then

$$\begin{aligned}\fint_{B_r(x_0)} \left|\eta u - \overline{(\eta u)}_{B_r(x_0)}\right| dx \le &\fint_{B_r(x_0)} \left|\eta(u - \bar{u}_{B_r(x_0)})\right| dx \\ &+ \fint_{B_r(x_0)} \left|\eta \bar{u}_{B_r(x_0)} - \overline{(\eta u)}_{B_r(x_0)}\right| dx.\end{aligned} \tag{2.7}$$

Clearly

$$\fint_{B_r(x_0)} \left|\eta(u - \bar{u}_{B_r(x_0)})\right| dx \le \|\eta\|_{L^\infty(\Omega)}[u]_{\mathrm{BMO}(\Omega)}. \tag{2.8}$$

Moreover,

$$\begin{aligned}
&\fint_{B_r(x_0)} \left|\eta \bar{u}_{B_r(x_0)} - \overline{(\eta u)}_{B_r(x_0)}\right| dx \\
&\quad = \fint_{B_r(x_0)} \left| \fint_{B_r(x_0)} (\eta(x) - \eta(y)) u(y)\, dy \right| dx \\
&\quad \le \|\nabla \eta\|_{L^\infty(\Omega)} \fint_{B_r(x_0)} \fint_{B_r(x_0)} |x-y||u(y)|\, dy\, dx \\
&\quad \le 2r \|\nabla \eta\|_{L^\infty(\Omega)} \left(\fint_{B_r(x_0)} |u|^m\, dx \right)^{1/m} \\
&\quad = 2\omega_m^{-1/m} \|\nabla \eta\|_{L^\infty(\Omega)} \left(\int_{B_r(x_0)} |u|^m\, dx \right)^{1/m}. \qquad (2.9)
\end{aligned}$$

It follows from (2.6)–(2.9) that

$$[v]_{\mathrm{BMO}(\Omega')} \le C_2 [u]_{\mathrm{BMO}(\Omega)}$$

for a constant C_2 that depends only on m, Ω, and η. For any ball $B_r(x_0) \subset \Omega$ that is neither contained in Ω' nor in $\Omega \backslash \operatorname{supp} \eta$, we have $r \ge C_3$ for a constant C_3 that depends only on Ω and η (and the choice of Ω'). Thus

$$\fint_{B_r(x_0)} |v - \bar{v}_{B_r(x_0)}|\, dx \le C_4 \|u\|_{L^m(\Omega')} \le C_5 [u]_{\mathrm{BMO}(\Omega)},$$

where C_4 and C_5 are also constants that depend only on m, Ω, and η. □

We also consider the convolution of a function $u \in \mathrm{BMO}(\mathbb{R}^m)$ with a Dirac sequence $\{\phi_s\}_{s>0}$. It is not true in general that we have convergence of $\phi_s * u$ to u in the BMO-sense as $s \to 0$, but at least the following holds.

Lemma 2.7 *Suppose $\phi \in C_0^\infty(B_1(0))$ with $\phi \ge 0$ and*

$$\int_{B_1(0)} \phi\, dx = 1.$$

For $s > 0$, write $\phi_s(x) = s^{-m}\phi(x/s)$. If $u \in \mathrm{BMO}(\mathbb{R}^m)$, then $\phi_s * u \in \mathrm{BMO}(\mathbb{R}^m)$ and

$$[\phi_s * u]_{\mathrm{BMO}(\mathbb{R}^m)} \le C_0 [u]_{\mathrm{BMO}(\mathbb{R}^m)}$$

for all $s > 0$, where C_0 is a constant that depends only on m and ϕ (but not on s).

Proof. We first prove the inequality for $s = 1$. Suppose $B_r(x_0) \subset \mathbb{R}^m$. If $r \le 1$, we use the fact that

$$\begin{aligned}\nabla(\phi * u)(x) &= \int_{B_1(x)} \nabla\phi(x-y)u(y)\,dy \\ &= \int_{B_1(x)} \nabla\phi(x-y)(u(y) - \bar{u}_{B_1(x)})\,dy\end{aligned}$$

and hence

$$\|\nabla(\phi * u)\|_{L^\infty(\mathbb{R}^m)} \le \|\nabla\phi\|_{L^\infty(B_1(0))}[u]_{\mathrm{BMO}(\mathbb{R}^m)}.$$

We conclude that

$$\operatorname*{osc}_{B_r(x_0)}(\phi * u) \le 2\|\nabla\phi\|_{L^\infty(B_1(0))}[u]_{\mathrm{BMO}(\mathbb{R}^m)}.$$

Hence

$$⨍_{B_r(x_0)} |\phi * u - \overline{(\phi * u)}_{B_r(x_0)}|\,dx \le 2\|\nabla\phi\|_{L^\infty(B_1(0))}[u]_{\mathrm{BMO}(\mathbb{R}^m)}.$$

On the other hand, if $r \ge 1$, we observe that

$$\begin{aligned}&⨍_{B_r(x_0)} |\phi * u - \bar{u}_{B_{2r}(x_0)}|\,dx \\ &\qquad = ⨍_{B_r(x_0)} \left| \int_{\mathbb{R}^m} \phi(y)(u(x-y) - \bar{u}_{B_{2r}(x_0)})\,dy \right| dx \\ &\qquad \le \int_{\mathbb{R}^m} \phi(y) ⨍_{B_r(x_0-y)} |u(x) - \bar{u}_{B_{2r}(x_0)}|\,dx\,dy \\ &\qquad \le 2^m ⨍_{B_{2r}(x_0)} |u - \bar{u}_{B_{2r}(x_0)}|\,dx \le 2^m [u]_{\mathrm{BMO}(\mathbb{R}^m)}.\end{aligned}$$

Analogously,

$$\begin{aligned}\left|\overline{(\phi * u)}_{B_r(x_0)} - \bar{u}_{B_{2r}(x_0)}\right| &= \left| ⨍_{B_r(x_0)} \int_{\mathbb{R}^m} \phi(y)(u(x-y) - \bar{u}_{B_{2r}(x_0)})\,dy\,dx \right| \\ &\le 2^m [u]_{\mathrm{BMO}(\mathbb{R}^m)}.\end{aligned}$$

Hence

$$
\begin{aligned}
&\fint_{B_r(x_0)} \left|\phi * u - \overline{(\phi * u)}_{B_r(x_0)}\right| dx \\
&\qquad \le \fint_{B_r(x_0)} |\phi * u - \bar{u}_{B_{2r}(x_0)}| \, dx + \left|\overline{(\phi * u)}_{B_r(x_0)} - \bar{u}_{B_{2r}(x_0)}\right| \\
&\qquad \le 2^{m+1}[u]_{\mathrm{BMO}(\mathbb{R}^m)}.
\end{aligned}
$$

This proves the claim of the lemma for $s = 1$.

For any other $s > 0$, we have

$$
\begin{aligned}
(\phi_s * u)(x) = s^{-m} \int_{\mathbb{R}^m} \phi(y/s) u(x - y) \, dy &= \int_{\mathbb{R}^m} \phi(y) u(x - sy) \, dy \\
&= (\phi * \tilde{u}_s)(x/s),
\end{aligned}
$$

where $\tilde{u}_s(x) = u(sx)$. Since $[\,\cdot\,]_{\mathrm{BMO}(\mathbb{R}^m)}$ is invariant under scaling, it follows that

$$
[\phi_s * u]_{\mathrm{BMO}(\mathbb{R}^m)} = [\phi * \tilde{u}_s]_{\mathrm{BMO}(\mathbb{R}^m)} \le C_0[\tilde{u}_s]_{\mathrm{BMO}(\mathbb{R}^m)} = C_0[u]_{\mathrm{BMO}(\mathbb{R}^m)},
$$

and the proof is complete. □

2.4 The Hardy Space $\mathcal{H}^1(\mathbb{R}^m)$

It is important for many of the proofs in this book to have a characterization of the dual space of $\mathrm{BMO}(\mathbb{R}^m)$. According to a result of C. Fefferman and E. M. Stein [1972], this dual space can be identified with the Hardy space $\mathcal{H}^1(\mathbb{R}^m)$. For this reason we give now a brief discussion of the space $\mathcal{H}^1(\mathbb{R}^m)$. It is not intended to give a complete picture of the properties of this space, but concentrates on the properties that are important for our purpose (that is, above all, the duality with $\mathrm{BMO}(\mathbb{R}^m)$). We consider $\mathcal{H}^1(\mathbb{R}^m)$ simply as a tool here. As a matter of fact, even the definition we give is intended to make the proofs as simple as possible. Other (equivalent) definitions could be given that might be more appropriate when the space is studied in its own right. The reader who is interested in a more thorough discussion can find it, e. g., in a book by E. M. Stein [1993].

Let $\mathcal{F}$ be the set of all $\phi \in C_0^\infty(B_1(0))$ such that

$$
\sup_{B_1(0)} |\nabla \phi| \le 1.
$$

For $s > 0$, we write $\phi_s(x) = s^{-m}\phi(x/s)$ whenever $\phi \in \mathcal{F}$. If u is a given distribution on $\mathbb{R}^m$, we consider the "maximal function" $\mathcal{M}u$, which is defined by

$$\mathcal{M}u(x) = \sup_{\phi\in\mathcal{F}} \sup_{s>0} |(\phi_s * u)(x)|, \quad x \in \mathbb{R}^m.$$

Definition 2.2 The *Hardy space* $\mathcal{H}^1(\mathbb{R}^m)$ consists of all distributions u on $\mathbb{R}^m$ such that

$$\|u\|_{\mathcal{H}^1(\mathbb{R}^m)} = \|\mathcal{M}u\|_{L^1(\mathbb{R}^m)} < \infty.$$

A possible way to identify this space with the dual space of $\mathrm{BMO}(\mathbb{R}^m)$ is to prove an "atomic decomposition" property of the following kind. This result was first proved by R. Coifman [1974] (for $m = 1$) and by R. H. Latter [1978]. The arguments we use here are from E. M. Stein [1993].

Proposition 2.1 *There exists a constant C, depending only on m, such that the following holds true. Suppose $u \in \mathcal{H}^1(\mathbb{R}^m) \cap L^1(\mathbb{R}^m)$. Then there exist a sequence of balls $B_{r_l}(x_l)$ and a sequence of functions $u_l \in L^\infty(\mathbb{R}^m)$ satisfying* $\operatorname{supp} u_l \subset B_{r_l}(x_l)$ *and*

$$\int_{\mathbb{R}^m} u_l \, dx = 0$$

for each l, such that

$$u = \sum_{l=1}^{\infty} u_l$$

(in the sense of pointwise convergence almost everywhere) and

$$\sum_{l=1}^{\infty} r_l^m \|u_l\|_{L^\infty(\mathbb{R}^m)} \le C\|u\|_{\mathcal{H}^1(\mathbb{R}^m)}.$$

Remark 2.4 *The convergence almost everywhere can be replaced by convergence in the $\mathcal{H}^1$-sense. The result of Proposition 2.1, however, is sufficient for our purpose.*

Proof. Consider the super-level sets

$$S_k = \left\{x \in \mathbb{R}^m : \ \mathcal{M}u(x) > 2^k\right\}, \quad k \in \mathbb{Z}.$$

To simplify the notation, we assume in the following that every S_k is non-empty. The arguments we use, however, work also in general.

It is readily verified that the sets S_k are open. Thus for every k there exists a collection of pairwise disjoint cubes $Q_i^k = Q_{r_i^k}(x_i^k)$, $i \in \mathbb{N}$, such that

(1) almost all of S_k is contained in the union

$$\bigcup_{i=1}^{\infty} Q_i^k,$$

(2) the inequality

$$4\sqrt{m}r_i^k \leq \operatorname{dist}(x_i^k, \mathbb{R}^m \backslash S_k) < 16\sqrt{m}r_i^k$$

holds, and

(3) there exists a number C_0, depending only on m, such that every point of S_k is contained in at most C_0 of the cubes $Q_{2r_i^k}(x_i^k)$.

For instance, the following construction (following ideas of H. Whitney) provides the desired decomposition of S_k: Let $i_0 \in \mathbb{Z}$ be the smallest number such that no cube of side length 2^{i_0+1} is contained in S_k. (Such a number exists, because the measure of S_k is finite.) Consider the set of all cubes with side length 2^{i_0} and vertices on the lattice $2^{i_0}\mathbb{Z}^m$. Choose every cube $Q_r(x)$ of this collection such that

$$4\sqrt{m}r \leq \operatorname{dist}(x, \mathbb{R}^m \backslash S_k). \tag{2.10}$$

Divide all other cubes in the set into 2^m smaller cubes of side length 2^{i_0-1} by bisection of the edges and choose again all the cubes that satisfy (2.10). Repeat this process indefinitely. The collection of all cubes that have been chosen in one of the steps has the desired properties.

Choose a function $\eta \in C_0^\infty(Q_2(0))$ with $0 \leq \eta \leq 1$ and $\eta \equiv 1$ in $Q_1(0)$. For each $k \in \mathbb{Z}$ and each $i \in \mathbb{N}$, set

$$\eta_i^k(x) = \eta((x - x_i^k)/r_i^k)$$

and

$$\xi_i^k(x) = \frac{\eta_i^k(x)}{\sum_{j=1}^{\infty} \eta_j^k(x)}.$$

Thus we obtain a partition of unity $\{\xi_i^k\}_{i \in \mathbb{N}}$ on S_k for each k. We have

$$\operatorname{supp} \xi_i^k \subset Q_{2r_i^k}(x_i^k)$$

and

$$\sup_{\mathbb{R}^m} |\nabla \xi_i^k| \le \frac{C_1}{r_i^k}$$

for a constant C_1 that depends only on m and the choice of η. Set now

$$\gamma_i^k = \frac{\int_{\mathbb{R}^m} u\xi_i^k \, dx}{\int_{\mathbb{R}^m} \xi_i^k \, dx}$$

and

$$c_{ij}^k = \frac{\int_{\mathbb{R}^m} (u - \gamma_j^{k+1})\xi_i^k \xi_j^{k+1} \, dx}{\int_{\mathbb{R}^m} \xi_j^{k+1} \, dx}.$$

For any fixed $k \in \mathbb{Z}$ and $i \in \mathbb{N}$ there exists a point $x_0 \in \mathbb{R}^m \backslash S_k$ such that

$$|x_i^k - x_0| \le 16\sqrt{m} r_i^k,$$

by the property (2) above. Thus for $s_i^k = 32\sqrt{m} r_i^k$, we have

$$\int_{\mathbb{R}^m} u\xi_i^k \, dx = C_2 (r_i^k)^m (\phi_{s_i^k} * u)(x_0)$$

for some $\phi \in \mathcal{F}$ and a constant C_2 that depends only on m and η. It follows that

$$\left| \int_{\mathbb{R}^m} u\xi_i^k \, dx \right| \le C_2 2^k (r_i^k)^m,$$

and thus

$$|\gamma_i^k| \le C_3 2^k$$

for another constant C_3 that depends only on m and η. A similar argument gives

$$|c_{ij}^k| \le C_4 2^k,$$

where C_4 is a constant that depends only on m and η.

Define

$$a_i^k = (u - \gamma_i^k)\xi_i^k - \sum_{j=1}^{\infty} (u - \gamma_j^{k+1})\xi_j^{k+1}\xi_i^k + \sum_{j=1}^{\infty} c_{ij}^k \xi_j^{k+1}.$$

We now examine the properties of the functions a_i^k. Note first that for $x \in S_{k+1}$, we have

$$|a_i^k(x)| = \left| \gamma_i^k \xi_i^k(x) - \sum_{j=1}^{\infty} \gamma_j^{k+1} \xi_j^{k+1}(x) \xi_i^k(x) - \sum_{j=1}^{\infty} c_{ij}^k \xi_j^{k+1}(x) \right| \leq C_5 2^k$$

for a constant C_5 that depends only on m and η. The terms containing u cancel because

$$\sum_{j=1}^{\infty} \xi_j^{k+1}(x) = 1.$$

For $x \notin S_{k+1}$, on the other hand, the functions ξ_j^{k+1} vanish; hence

$$|a_i^k(x)| = |(u(x) - \gamma_i^k) \xi_i^k(x)| \leq (C_3 + 1) 2^k.$$

We conclude that

$$\|a_i^k\|_{L^\infty(\mathbb{R}^m)} \leq (C_3 + C_5 + 1) 2^k.$$

Moreover, the support of a_i^k is contained in the union of $Q_{2r_i^k}(x_i^k)$ with all cubes $Q_{2r_j^{k+1}}(x_j^{k+1})$ such that

$$Q_{2r_i^k}(x_i^k) \cap Q_{2r_j^{k+1}}(x_j^{k+1}) \neq \emptyset.$$

This union is contained in a ball around x_i^k of radius Cr_i^k. We also have the inequality

$$\sum_{k=-\infty}^{\infty} \sum_{i=1}^{\infty} \|a_i^k\|_{L^\infty(\mathbb{R}^m)} (r_i^k)^m \leq C_6 \sum_{k=-\infty}^{\infty} 2^k |S_k| \leq C_7 \|u\|_{\mathcal{H}^1(\mathbb{R}^m)}$$

for constants C_6 and C_7 that depend only on m and η. Moreover,

$$\int_{\mathbb{R}^m} a_i^k \, dx = 0$$

by the definitions of γ_i^k and c_{ij}^k.

Next we want to prove that

$$u = \sum_{k=-\infty}^{\infty} \sum_{i=1}^{\infty} a_i^k.$$

To this end, define first

$$b_k = \sum_{i=1}^{\infty}(u - \gamma_i^k)\xi_i^k$$

and

$$u_k = u - b_k.$$

Note that $b_k = 0$ in $\mathbb{R}^m \backslash S_k$. Hence $b_k \to 0$ as $k \to \infty$ almost everywhere. Moreover, we have $|u_k(x)| \leq 2^k$ for almost every $x \in \mathbb{R}^m \backslash S_k$ and

$$|u_k(x)| \leq \sum_{i=1}^{\infty} |\gamma_i^k \xi_i^k(x)| \leq C_8 2^k$$

for $x \in S_k$, where C_8 is a constant that depends on m and η. In particular $u_k \to 0$ as $k \to -\infty$ almost everywhere. Hence

$$u_{k_0} = \sum_{k=-\infty}^{k_0-1} (u_{k+1} - u_k) = \sum_{k=-\infty}^{k_0-1} (b_k - b_{k+1}).$$

That is,

$$u = \sum_{k=-\infty}^{\infty} (b_k - b_{k+1}).$$

Note further that

$$\sum_{i=1}^{\infty} c_{ij}^k = 0$$

for all $k \in \mathbb{Z}$ and all $j \in \mathbb{N}$ by the definitions of c_{ij}^k and γ_i^k. It follows that

$$\sum_{i=1}^{\infty} a_i^k = b_k - \sum_{j=1}^{\infty}(u - \gamma_j^{k+1})\xi_j^{k+1} = b_k - b_{k+1}.$$

Thus

$$\sum_{k=-\infty}^{\infty} \sum_{i=1}^{\infty} a_i^k = \sum_{k=-\infty}^{\infty} (b_k - b_{k+1}) = u.$$

To obtain a sequence with the desired properties, it now suffices to relabel the functions a_i^k. □

If we have a function $u \in \mathrm{BMO}(\mathbb{R}^m)$ and an "atom" like in Proposition 2.1, that is, a function $v \in L^\infty(\mathbb{R}^m)$ with $\operatorname{supp} v \subset B_r(x_0)$ and

$$\int_{\mathbb{R}^m} v\, dx = 0,$$

then clearly

$$\int_{\mathbb{R}^m} uv\, dx \leq \omega_m r^m [u]_{\mathrm{BMO}(\mathbb{R}^m)} \|v\|_{L^\infty(\mathbb{R}^m)}.$$

In order to estimate such an integral for a general $v \in \mathcal{H}^1(\mathbb{R}^m) \cap L^1(\mathbb{R}^m)$, we need to be able to exchange the integral and the summation in the atomic decomposition of v. This is possible, e. g., if $u \in L^\infty(\mathbb{R}^m)$.

Theorem 2.3 *Suppose $u \in L^\infty(\mathbb{R}^m)$ and $v \in \mathcal{H}^1(\mathbb{R}^m) \cap L^1(\mathbb{R}^m)$. Then*

$$\int_{\mathbb{R}^m} uv\, dx \leq C[u]_{\mathrm{BMO}(\mathbb{R}^m)} \|v\|_{\mathcal{H}^1(\mathbb{R}^m)}$$

for a constant C that depends only on m.

Proof. Let

$$v = \sum_{l=1}^{\infty} v_l$$

be the atomic decomposition of Proposition 2.1. That is, the functions $v_l \in L^\infty(\mathbb{R}^m)$ have the properties

$$\int_{\mathbb{R}^m} v_l\, dx = 0$$

and $\operatorname{supp} v_l \subset B_{r_l}(x_l)$ for certain balls satisfying

$$\sum_{l=1}^{\infty} r_l^m \|v_l\|_{L^\infty(\mathbb{R}^m)} \leq C\|v\|_{\mathcal{H}^1(\mathbb{R}^m)}.$$

Then the function

$$G = \|u\|_{L^\infty(\mathbb{R}^m)} \sum_{l=1}^{\infty} |v_l|$$

is in $L^1(\mathbb{R}^m)$. Setting

$$G_k = \sum_{l=1}^{k} uv_l,$$

we have $|G_k| \leq G$ and $G_k \to uv$ as $k \to \infty$ almost everywhere. By Lebesgue's convergence theorem,

$$\int_{\mathbb{R}^m} uv\,dx = \lim_{k\to\infty} \int_{\mathbb{R}^m} G_k\,dx = \sum_{l=1}^{\infty} \int_{\mathbb{R}^m} uv_l\,dx$$

$$\leq \omega_m \sum_{l=1}^{\infty} r_l^m [u]_{\mathrm{BMO}(\mathbb{R}^m)} \|v_l\|_{L^\infty(\mathbb{R}^m)} \leq C[u]_{\mathrm{BMO}(\mathbb{R}^m)} \|v\|_{\mathcal{H}^1(\mathbb{R}^m)},$$

as claimed. □

Remark 2.5 *For general $u \in \mathrm{BMO}(\mathbb{R}^m)$ and $v \in \mathcal{H}^1(\mathbb{R}^m)$, the integral*

$$\int_{\mathbb{R}^m} uv\,dx$$

may be undefined. The inequality of Theorem 2.3 holds, however, for every $u \in \mathrm{BMO}(\mathbb{R}^m)$ and every finite sum of atoms like in Proposition 2.1. The space of such finite sums is dense in $\mathcal{H}^1(\mathbb{R}^m)$, therefore the bilinear functional defined thus can be extended continuously to $\mathrm{BMO}(\mathbb{R}^m) \times \mathcal{H}^1(\mathbb{R}^m)$. Using this extension, the dual space of $\mathrm{BMO}(\mathbb{R}^m)$ can be identified with $\mathcal{H}^1(\mathbb{R}^m)$. (Details about these arguments can be found in Chapter IV of E. M. Stein [1993].) For our purpose, however, Theorem 2.3 is sufficient.

Finally, if we want to apply the previous theorem successfully, we need to be able to estimate the $\mathcal{H}^1$-norm of certain expressions. In particular, we consider the scalar product of a gradient vector field with a divergence free vector field. The following compensated compactness theorem is a version of a result of R. Coifman, P. L. Lions, Y. Meyer, and S. Semmes [1993]. Related results have been proved by S. Chanillo [1991], M. Feldman [1994], and others.

Theorem 2.4 *Let $\delta > 0$ and $p \in [1+\delta, 1/\delta]$. If u is in the Sobolev space $W^{1,p}(\mathbb{R}^m)$ and $X \in L^{p/(p-1)}(\mathbb{R}^m;\mathbb{R}^m)$ satisfies $\operatorname{div} X = 0$ in the distribution sense, then the function $v = \nabla u \cdot X$ is in $\mathcal{H}^1(\mathbb{R}^m)$. Moreover, there exists a constant C_0, depending only on m and δ, such that*

$$\|v\|_{\mathcal{H}^1(\mathbb{R}^m)} \leq C_0 \|\nabla u\|_{L^p(\mathbb{R}^m)} \|X\|_{L^{p/(p-1)}(\mathbb{R}^m)}.$$

Proof. Suppose $\phi \in \mathcal{F}$. For any $x \in \mathbb{R}^m$ and any $s > 0$, we have

$$(\phi_s * v)(x) = \int_{B_s(x)} \phi_s(x-y) \nabla u(y) \cdot X(y)\,dy$$

$$= \int_{B_s(x)} \nabla\phi_s(x-y) \cdot X(y)\,(u(y) - \bar{u}_{B_s(x)})\,dy. \quad (2.11)$$

If $p < m$, choose a number $q \in (p, pm/(m-p))$. Otherwise choose $q > p$. We can do this such that

$$p - \frac{mq}{m+q} \geq \gamma \quad \text{and} \quad \frac{p}{p-1} - \frac{q}{q-1} \geq \gamma \tag{2.12}$$

for a constant $\gamma > 0$ that depends only on δ. Then

$$|(\phi_s * v)(x)| \\ \leq \frac{\omega_m}{s} \left(\fint_{B_s(x)} |u - \bar{u}_{B_s(x)}|^q \, dy \right)^{1/q} \left(\fint_{B_s(x)} |X|^{q/(q-1)} \, dy \right)^{(q-1)/q}.$$

By the Sobolev inequality, we have

$$\left(\fint_{B_s(x)} |u - \bar{u}_{B_s(x)}|^q \, dy \right)^{1/q} \leq C_1 s \left(\fint_{B_s(x)} |\nabla u|^{mq/(m+q)} \, dy \right)^{\frac{m+q}{mq}}$$

for a constant C_1 that depends only on m and q. Hence there exists another constant C_2, depending only on m and q, such that

$$|(\phi_s * v)(x)| \leq C_2 \left(\fint_{B_s(x)} |\nabla u|^{mq/(m+q)} \, dy \right)^{\frac{m+q}{mq}} \left(\fint_{B_s(x)} |X|^{q/(q-1)} \, dy \right)^{\frac{q-1}{q}}. \tag{2.13}$$

Thus, setting

$$(M(f))(x) = \sup_{s>0} \fint_{B_s(x)} f \, dy$$

for any non-negative measurable function f on $\mathbb{R}^m$ and any $x \in \mathbb{R}^m$, we have

$$\sup_{\phi \in \mathcal{F}} \sup_{s>0} |\phi_s * v| \leq C_3 M(|\nabla u|^{mq/(m+q)})^{(m+q)/(mq)} M(|X|^{q/(q-1)})^{(q-1)/q}$$

for a constant C_3 that depends only on m and q. Because of (2.12) and by the Hardy-Littlewood maximal theorem, we have thus

$$\left\| \sup_{\phi \in \mathcal{F}} \sup_{s>0} |\phi_s * v| \right\|_{L^1(\mathbb{R}^m)} \leq C_4 \|\nabla u\|_{L^p(\mathbb{R}^m)} \|X\|_{L^{p/(p-1)}(\mathbb{R}^m)},$$

where C_4 is a constant that depends only on m, p, and δ. □

The key observation in the proof of Theorem 2.4 is that a product of the form $v = \nabla u \cdot X$ is a divergence if $\operatorname{div} X = 0$, namely $v = \operatorname{div}(uX)$. We can prove similar inequalities for the divergence of a product of this kind even if we do not have the condition $\operatorname{div} X = 0$, although these estimates are not quite as good as Theorem 2.4.

Theorem 2.5 *For any $\delta > 0$ there exists a number C_0, depending only on m and δ, with the following property: Suppose $p, q \in [1+\delta, 1/\delta]$ and $\frac{1}{p} + \frac{1}{p'} = \frac{1}{q} + \frac{1}{q'} = 1$. If $u \in W^{1,p}(\mathbb{R}^m) \cap L^{q'}(\mathbb{R}^m)$ and $X \in W^{1,q}(\mathbb{R}^m; \mathbb{R}^m) \cap L^{p'}(\mathbb{R}^m; \mathbb{R}^m)$, then the function $v = \operatorname{div}(uX)$ is in $\mathcal{H}^1(\mathbb{R}^m)$ with*

$$\|v\|_{\mathcal{H}^1(\mathbb{R}^m)} \le C_0 \left(\|\nabla u\|_{L^p(\mathbb{R}^m)} \|X\|_{L^{p'}(\mathbb{R}^m)} + \|u\|_{L^{q'}(\mathbb{R}^m)} \|\nabla X\|_{L^q(\mathbb{R}^m)} \right).$$

Proof. Let $\phi \in \mathcal{F}$ and $s > 0$. For any fixed $x \in \mathbb{R}^m$ we have

$$\begin{aligned}
(\phi_s * v)(x) &= \int_{B_s(x)} \phi_s(x-y) \operatorname{div}(u(y)(X(y) - \bar{X}_{B_s(x)}))\, dy \\
&\quad + \int_{B_s(x)} \phi_s(x-y) \nabla(u(y) - \bar{u}_{B_s(x)}) \cdot \bar{X}_{B_s(x)}\, dy \\
&= \int_{B_s(x)} \nabla\phi_s(x-y) \cdot (X(y) - \bar{X}_{B_s(x)})\, u(y)\, dy \\
&\quad + \int_{B_s(x)} \nabla\phi_s(x-y) \cdot \bar{X}_{B_s(x)}\, (u(y) - \bar{u}_{B_s(x)})\, dy.
\end{aligned}$$

Choose a number α such that

$$\frac{m\alpha}{m+\alpha} < q < \alpha$$

and set

$$\alpha' = \frac{\alpha}{\alpha - 1}, \quad \beta = \frac{m\alpha}{m+\alpha}.$$

We can do this in a way such that $q - \beta \ge \gamma$ and $q' - \alpha' \ge \gamma$ for a number $\gamma > 0$ that depends only on m and δ. Using the Hölder inequality and the Sobolev inequality, we find

$$\begin{aligned}
&\left| \int_{B_s(x)} \nabla\phi_s(x-y) \cdot (X(y) - \bar{X}_{B_s(x)})\, u(y)\, dy \right| \\
&\quad \le \frac{\omega_m}{s} \left(⨍_{B_s(x)} |X - \bar{X}_{B_s(x)}|^\alpha\, dy \right)^{1/\alpha} \left(⨍_{B_s(x)} |u|^{\alpha'}\, dy \right)^{1/\alpha'} \\
&\quad \le C_1 \left(⨍_{B_s(x)} |\nabla X|^\beta\, dy \right)^{1/\beta} \left(⨍_{B_s(x)} |u|^{\alpha'}\, dy \right)^{1/\alpha'}
\end{aligned}$$

for a constant C_1 that depends only on m and δ. We also have

$$\begin{aligned}\left|\int_{B_s(x)} \nabla\phi_s(x-y)\cdot \bar{X}_{B_s(x)}\,(u(y)-\bar{u}_{B_s(x)})\,dy\right| \\ &\leq \frac{\omega_m}{s} \fint_{B_s(x)} |X|\,dy \fint_{B_s(x)} |u-\bar{u}_{B_s(x)}|\,dy \\ &\leq C \fint_{B_s(x)} |X|\,dy \fint_{B_s(x)} |\nabla u|\,dy.\end{aligned}$$

Setting again

$$(M(f))(x) = \sup_{s>0} \fint_{B_s(x)} f\,dy,$$

we conclude that

$$\sup_{\phi\in\mathcal{F}} \sup_{s>0} |\phi_s * v| \leq C_1 M(|\nabla X|^\beta)^{1/\beta} M(|u|^{\alpha'})^{1/\alpha'} + CM(|X|)M(|\nabla u|).$$

The claim now follows from the Hardy-Littlewood maximal theorem, like in the proof of Theorem 2.4. □

2.5 A Gagliardo-Nirenberg Type Inequality

The following is a well-known inequality: If $p > 1$ and $u \in W^{2,p}(\mathbb{R}^m) \cap L^\infty(\mathbb{R}^m)$, then $\nabla u \in L^{2p}(\mathbb{R}^m;\mathbb{R}^m)$ and

$$\|\nabla u\|^2_{L^{2p}(\mathbb{R}^m)} \leq C_0 \|u\|_{L^\infty(\mathbb{R}^m)} \|\nabla^2 u\|_{L^p(\mathbb{R}^m)} \tag{2.14}$$

for a constant C_0 that depends only on m and p. Many inequalities of this type, including (2.14), have been proved by L. Nirenberg [1959].

We will need an improvement of (2.14) for some of the results in the subsequent chapters, one that involves the BMO-seminorm instead of the L^∞-norm. A possible proof of such an inequality has been sketched by D. R. Adams and M. Frazier [1992]. We give a different proof here, based on the BMO-$\mathcal{H}^1$-duality and the results of the previous section.

Proposition 2.2 *For any $\delta > 0$ there exists a constant C_0, depending only on m and δ, such that*

$$\|\nabla u\|^2_{L^{2p}(\mathbb{R}^m)} \leq C_0 [u]_{\mathrm{BMO}(\mathbb{R}^m)} \|\nabla^2 u\|_{L^p(\mathbb{R}^m)} \tag{2.15}$$

for any $p \in [1+\delta, 1/\delta]$ and any $u \in W^{2,p}(\mathbb{R}^m) \cap \mathrm{BMO}(\mathbb{R}^m)$.

Proof. Choose a function $\phi \in C_0^\infty(B_1(0))$ with

$$\int_{B_1(0)} \phi \, dx = 1.$$

For $s > 0$, set $\phi_s(x) = s^{-m}\phi(x/s)$ and

$$u_s = \phi_s * u.$$

By Lemma 2.7, we have

$$[u_s]_{\mathrm{BMO}(\mathbb{R}^m)} \leq C_1[u]_{\mathrm{BMO}(\mathbb{R}^m)}$$

for a constant C_1 that depends only on m and ϕ. Moreover,

$$\lim_{s \searrow 0} \|\nabla^2 u_s\|_{L^p(\mathbb{R}^m)} = \|\nabla^2 u\|_{L^p(\mathbb{R}^m)}.$$

If we can prove (2.15) for all smooth maps, it follows that $\nabla u \in L^{2p}(\mathbb{R}^m; \mathbb{R}^m)$ and thus

$$\lim_{s \searrow 0} \|\nabla u_s\|_{L^{2p}(\mathbb{R}^m)} = \|\nabla u\|_{L^{2p}(\mathbb{R}^m)}.$$

Hence (2.15) holds also for u (possibly with a different constant). Thus it suffices to consider smooth maps. By a similar approximation argument, using Lemma 2.6, we see that we may even assume $u \in C_0^\infty(\mathbb{R}^m)$.

Now we have

$$\int_{\mathbb{R}^m} |\nabla u|^{2p} \, dx = -\int_{\mathbb{R}^m} u \operatorname{div}(|\nabla u|^{2p-2}\nabla u) \, dx.$$

We apply Theorem 2.5 for $f = |\nabla u|^{2p-2}$ and $X = \nabla u$. Since we have

$$\|\nabla X\|_{L^p(\mathbb{R}^m)} \leq \|\nabla^2 u\|_{L^p(\mathbb{R}^m)}$$

and

$$\|\nabla f\|_{L^{2p/(2p-1)}(\mathbb{R}^m)} \leq (2p-2)\|\nabla u\|_{L^{2p}(\mathbb{R}^m)}^{2p-3} \|\nabla^2 u\|_{L^p(\mathbb{R}^m)},$$

it follows that

$$\|\operatorname{div}(|\nabla u|^{2p-2}\nabla u)\|_{\mathcal{H}^1(\mathbb{R}^m)} \leq C_2 \|\nabla u\|_{L^{2p}(\mathbb{R}^m)}^{2p-2} \|\nabla^2 u\|_{L^p(\mathbb{R}^m)}$$

for a constant C_2 that depends only on m and δ. Using Theorem 2.3, we find that

$$\|\nabla u\|_{L^{2p}(\mathbb{R}^m)}^{2p} \leq C_3 [u]_{\mathrm{BMO}(\mathbb{R}^m)} \|\nabla u\|_{L^{2p}(\mathbb{R}^m)}^{2p-2} \|\nabla^2 u\|_{L^p(\mathbb{R}^m)}$$

for a constant C_3 that depends on m and δ, and (2.15) follows. □

2.6 A Gehring Type Lemma

It was first observed by F. W. Gehring [1973] that reverse Hölder inequalities can be used to improve the integrability of a function. These ideas were improved by M. Giaquinta and G. Modica [1979]. The resulting lemma is a tool that is useful for the study of many variational problems, among them the problem of harmonic maps. The arguments we present in this section are basically the same as in the works of M. Giaquinta and G. Modica [1979] and F. W. Gehring [1973] (see also M. Giaquinta [1983]).

First, we need some preparation. For a given non-negative function $u : \Omega \to [0, \infty)$ on a measurable set Ω and for $s > 0$, we write, similarly as before,

$$S(u, s) = \{x \in \Omega : \ u(x) > s\}.$$

Then we have

$$\int_\Omega u^p \, dx = \int_0^\infty |S(u^p, s)| \, ds = p \int_0^\infty s^{p-1} |S(u, s)| \, ds$$

for any $p \geq 1$ by Fubini's theorem and the substitution formula. If $p > 1$ and $u \in L^p(\Omega)$, an integration by parts yields further

$$\begin{aligned}\int_\Omega u^p \, dx &= p \lim_{s_1 \searrow 0} \lim_{s_2 \to \infty} \int_{s_1}^{s_2} s^{p-1} |S(u, s)| \, ds \\ &= p \lim_{s_1 \searrow 0} \lim_{s_2 \to \infty} \left(s_1^{p-1} \int_{s_1}^\infty |S(u, s)| \, ds - s_2^{p-1} \int_{s_2}^\infty |S(u, s)| \, ds \right. \\ &\qquad \left. + (p-1) \int_{s_1}^{s_2} s^{p-2} \int_s^\infty |S(u, t)| \, dt \, ds \right).\end{aligned}$$

We have

$$s^{p-1} \int_s^\infty |S(u, t)| \, dt \leq s^{p-1} \int_{S(u,s)} u \, dx \leq (s^p |S(u, s)|)^{(p-1)/p} \|u\|_{L^p(\Omega)}$$

by Hölder's inequality. From

$$\int_0^\infty s^{p-1} |S(u, s)| \, ds < \infty,$$

it follows that

$$\liminf_{s_1 \searrow 0} (s_1^p |S(u, s_1)|) = \liminf_{s_2 \to \infty} (s_2^p |S(u, s_2)|) = 0.$$

Hence

$$\liminf_{s_1 \searrow 0} \left(s_1^{p-1} \int_{s_1}^{\infty} |S(u,s)|\, ds \right) = \liminf_{s_2 \to \infty} \left(s_2^{p-1} \int_{s_2}^{\infty} |S(u,s)|\, ds \right) = 0$$

and

$$\begin{aligned}
\int_\Omega u^p\, dx &= p(p-1) \int_0^\infty s^{p-2} \int_s^\infty |S(u,t)|\, dt\, ds \\
&= p(p-1) \int_0^\infty s^{p-2} \left(\int_{S(u,s)} u\, dx - s|S(u,s)| \right) ds \\
&= p(p-1) \int_0^\infty s^{p-2} \int_{S(u,s)} u\, dx\, ds - (p-1) \int_\Omega u^p\, dx.
\end{aligned}$$

Therefore,

$$\int_\Omega u^p\, dx = (p-1) \int_0^\infty s^{p-2} \int_{S(u,s)} u\, dx\, ds. \tag{2.16}$$

These facts play an essential role in the next lemma.

Lemma 2.8 *Suppose $p > 1$. Let $u, v \in L^p(\Omega)$ be two non-negative functions and $a > 0$ a number such that*

$$\int_{S(u,s)} u^p\, dx \le a \left(s^{p-1} \int_{S(u,s)} u\, dx + \int_{S(v,s)} v^p\, dx \right) \tag{2.17}$$

for every $s > 0$. Then for every number q satisfying

$$p \le q < p + \frac{p-1}{a},$$

the inequality

$$\int_\Omega u^q\, dx \le \frac{a}{p-1-a(q-p)} \int_\Omega v^q\, dx \tag{2.18}$$

holds.

Proof. Let $0 < s_1 < s_2 < \infty$ and define

$$I(q) = \int_{s_1}^{s_2} s^{q-2} \int_{S(u,s)} u\, dx\, ds, \quad q \ge p.$$

By an integration by parts, we find

$$\begin{aligned}
I(q) &= \int_{s_1}^{s_2} s^{q-p} s^{p-2} \int_{S(u,s)} u\,dx\,ds \\
&= s_1^{q-p} \int_{s_1}^{\infty} s^{p-2} \int_{S(u,s)} u\,dx\,ds - s_2^{q-p} \int_{s_2}^{\infty} s^{p-2} \int_{S(u,s)} u\,dx\,ds \\
&\quad + (q-p) \int_{s_1}^{s_2} s^{q-p-1} \int_{s}^{\infty} t^{p-2} \int_{S(u,t)} u\,dx\,dt\,ds \\
&\le \frac{s_1^{q-p}}{p-1} \int_{S(u,s_1)} u^p\,dx + \frac{q-p}{p-1} \int_{s_1}^{s_2} s^{q-p-1} \int_{S(u,s)} u^p\,dx\,ds.
\end{aligned}$$

In the last step, we have used (2.16). Now (2.17) implies

$$\begin{aligned}
I(q) &\le \frac{s_1^{q-p}}{p-1} \int_{S(u,s_1)} u^p\,dx \\
&\quad + \frac{(q-p)a}{p-1} \int_{s_1}^{s_2} \left(s^{q-2} \int_{S(u,s)} u\,dx + s^{q-p-1} \int_{S(v,s)} v^p\,dx \right) ds \\
&= \frac{s_1^{q-p}}{p-1} \int_{S(u,s_1)} u^p\,dx \\
&\quad + \frac{(q-p)a}{p-1} \left(I(q) + \int_{s_1}^{s_2} s^{q-p-1} \int_{S(v,s)} v^p\,dx\,ds \right).
\end{aligned}$$

We can assume $v \in L^q(\Omega)$, for inequality (2.18) is trivial otherwise. Then we have

$$\begin{aligned}
\int_{s_1}^{s_2} s^{q-p-1} \int_{S(v,s)} v^p\,dx\,ds &\le \int_{0}^{\infty} s^{q-p-1} \int_{S(v,s)} v^p\,dx\,ds \\
&= \frac{1}{p} \int_{0}^{\infty} s^{q/p-2} \int_{S(v^p,s)} v^p\,dx\,ds \\
&= \frac{1}{q-p} \int_{\Omega} v^q\,dx
\end{aligned}$$

by the substitution formula and (2.16). If

$$\frac{(q-p)a}{p-1} < 1,$$

we conclude tha+

$$I(q) \le \frac{1}{1 - \frac{(q-p)a}{p-1}} \left(\frac{s_1^{q-p}}{p-1} \int_{S(u,s_1)} u^p\,dx + \frac{a}{p-1} \int_{\Omega} v^q\,dx \right).$$

If we let s_1 converge to 0, the first term on the right hand side vanishes in the limit, and (2.18) follows. □

We work again with cubes with edges parallel to the coordinate axes for most of the rest of this section. We use the same notation as in Section 2.3. We also use the following decomposition of the unit cube $Q_1(0)$: Let $K_0 = Q_{1/2}(0)$ and

$$\begin{aligned} K_l &= Q_{1-2^{-l-1}}(0)\backslash Q_{1-2^{-l}}(0) \\ &= \left\{x \in Q_1(0) :\ 2^{-l-1} < \operatorname{dist}(x, \partial Q_1(0)) \le 2^{-l}\right\}, \quad l \in \mathbb{N}. \end{aligned}$$

Then we have

$$Q_1(0) = \bigcup_{l=0}^{\infty} K_l.$$

We will need the following modified version of the Calderón-Zygmund cube decomposition lemma.

Lemma 2.9 *Let u be an integrable, non-negative function on $Q_1(0)$. Suppose $s > 0$ is a number such that*

$$\fint_{Q_1(0)} u\,dx \le s. \tag{2.19}$$

Then there exists a finite or countably infinite collection $\{Q_i\}$ (where i is between 1 and $I \le \infty$) of pairwise disjoint cubes, such that

(1) each Q_i is contained in one of the sets $K_0, K_1, \ldots,$
(2) if $Q_i \subset K_l$, then

$$2^{lm}s < \fint_{Q_i} u\,dx \le 2^{(l+1)m}s,$$

and
(3) for every $l \ge 0$, the inequality $u \le 2^{lm}s$ holds almost everywhere in

$$K_l \backslash \bigcup_{i=1}^{I} Q_i.$$

Proof. The arguments are almost the same as for the classical Calderón-Zygmund cube decomposition. We construct the collection $Q_1, Q_2, \ldots$ as follows: We first decompose $Q_1(0)$ into 2^m congruent subcubes by bisection of the edges. None of the cubes we obtain in this first step is contained in one of the sets $K_0, K_1, \ldots,$ so we divide each of them again into 2^m cubes of side length $\frac{1}{2}$. (That gives a total of 2^{2m} cubes.) Now we collect all the cubes Q of this subdivision such that there exists some $l \geq 0$ with $Q \subset K_l$ and

$$\fint_Q u\,dx > 2^{lm}s.$$

All other cubes are further divided and this process is continued indefinitely.

The collection $\{Q_i\}$ finally consists of all cubes that have been chosen in one such step. Clearly this collection satisfies (1). Moreover, if $Q_i \subset K_l$, then

$$\fint_{Q_i} u\,dx > 2^{lm}s.$$

Denote by $\tilde{Q}$ the cube whose subdivision yielded Q_i. Obviously $\tilde{Q}$ was not chosen in the previous step, hence it satisfies either

$$\fint_{\tilde{Q}} u\,dx \leq 2^{lm}s$$

and thus

$$\fint_{Q_i} u\,dx \leq 2^{(l+1)m}s, \tag{2.20}$$

or $\tilde{Q} \not\subset K_l$. The latter can only be true if the side length of $\tilde{Q}$ is at least 2^{-l} (by the definition of K_l and because $Q_i \subset K_l$). The condition (2.19) then implies (2.20), too. Hence (2) is true. The property (3) is a consequence of Lebesgue's differentiation theorem, since for almost every $x \in K_l \backslash \bigcup_{i=1}^{I} Q_i$, we have a sequence of cubes P_j with $x \in P_j$ and $\lim_{j\to\infty} \operatorname{diam} P_j = 0$, such that

$$\fint_{P_j} u\,dx \leq 2^{ml}s.$$

This concludes the proof. □

We are now in a position to prove a version of the main lemma of this section. Another version, with balls instead of cubes, will follow at the end of the section.

Lemma 2.10 *Suppose $1 < p < q$ and $a > 0$. There exist three numbers $\bar{p} \in (p, q]$, $\theta > 0$, and $b > 0$, depending only on m, p, q, and a, such that* *the following holds: Let* $u \in L^p(Q_1(0))$ *and* $v \in L^q(Q_1(0))$ *be non-negative functions satisfying*

$$\fint_{Q_{r/2}(x_0)} u^p\,dx \le \theta \fint_{Q_r(x_0)} u^p\,dx + a\left(\fint_{Q_r(x_0)} u\,dx\right)^p + \fint_{Q_r(x_0)} v^p\,dx$$

for every cube $Q_r(x_0) \subset Q_1(0)$*. Then* $u \in L^{\bar{p}}(Q_{1/2}(0))$ *and*

$$\int_{Q_{1/2}(0)} u^{\bar{p}}\,dx \le b\left(\int_{Q_1(0)} u^p\,dx\right)^{\bar{p}/p} + b\int_{Q_1(0)} v^{\bar{p}}\,dx.$$

Proof. Define the functions f and g on $Q_1(0)$ by

$$f(x) = 2^{-lm/p}u(x) \quad \text{in } K_l$$

and

$$g(x) = 2^{-lm/p}v(x) \quad \text{in } K_l$$

for $l \ge 0$. Assume $c \ge 1$ is a given constant. We will determine its value later. Suppose t is a positive number such that

$$c^p t^p > \fint_{Q_1(0)} f^p\,dx. \tag{2.21}$$

Set $s = ct$. We apply Lemma 2.9 to the function u^p for the number s^p instead of s. Thus we find a collection $\{Q_{r_i}(x_i)\}$ (where i is between 1 and $I \le \infty$) of cubes satisfying $Q_{2r_i}(x_i) \subset Q_1(0)$ for every i, such that $f^p \le s^p$ almost everywhere in $Q_1(0) \setminus \bigcup_{i=1}^{I} Q_{r_i}(x_i)$ and

$$s^p \le \fint_{Q_{r_i}(x_i)} f^p\,dx \le 2^m s^p. \tag{2.22}$$

We set

$$S(f,s) = \{x \in Q_1(0) : \; f(x) > s\}$$

and

$$S(g,s) = \{x \in Q_1(0) : \; g(x) > s\}.$$

Then we have

$$S(f,s) \subset \bigcup_{i=1}^{I} Q_{r_i}(x_i).$$

Hence

$$\int_{S(f,s)} f^p\,dx \le \sum_{i=1}^{I} \int_{Q_{r_i}(x_i)} f^p\,dx \le 4^m s^p \sum_{i=1}^{I} r_i^m. \tag{2.23}$$

The first inequality in (2.22) and the conditions of the lemma imply that

$$\begin{aligned} s^p &\le \fint_{Q_{r_i}(x_i)} f^p\,dx \\ &\le 2^m\theta \fint_{Q_{2r_i}(x_i)} f^p\,dx + 2^m a\left(\fint_{Q_{2r_i}(x_i)} f\,dx\right)^p + 2^m \fint_{Q_{2r_i}(x_i)} g^p\,dx, \end{aligned}$$

hence

$$\begin{aligned} s r_i^m \le r_i^{m-m/p}&\left(\theta\int_{Q_{2r_i}(x_i)} f^p\,dx\right)^{1/p} + a^{1/p}\int_{Q_{2r_i}(x_i)} f\,dx \\ &+ r_i^{m-m/p}\left(\int_{Q_{2r_i}(x_i)} g^p\,dx\right)^{1/p}. \end{aligned}$$

Note that

$$\begin{aligned} a^{1/p}\int_{Q_{2r_i}(x_i)} f\,dx &\le a^{1/p}\left(\int_{S(f,t)\cap Q_{2r_i}(x_i)} f\,dx + 4^m r_i^m t\right) \\ &\le a^{1/p}\left(\int_{S(f,t)\cap Q_{2r_i}(x_i)} f\,dx + \frac{4^m r_i^m s}{c}\right). \end{aligned}$$

Moreover,

$$r_i^{m-m/p}\left(\theta\int_{Q_{2r_i}(x_i)} f^p\,dx\right)^{1/p} \le \theta^{1/p} s^{1-p}\int_{Q_{2r_i}(x_i)} f^p\,dx,$$

because of (2.22). We conclude that

$$\begin{aligned} r_i^{m-m/p}&\left(\theta\int_{Q_{2r_i}(x_i)} f^p\,dx\right)^{1/p} \\ &\le \theta^{1/p} s^{1-p}\left(\int_{S(f,s)\cap Q_{2r_i}(x_i)} f^p\,dx + 4^m r_i^m s^p\right). \end{aligned}$$

A similar reasoning gives

$$
\begin{aligned}
r_i^{m-m/p}&\left(\int_{Q_{2r_i}(x_i)} g^p\,dx\right)^{1/p}\\
&\le s^{1-p}\left(\int_{Q_{2r_i}(x_i)} f^p\,dx\right)^{1-1/p}\left(\int_{Q_{2r_i}(x_i)} g^p\,dx\right)^{1/p}\\
&\le s^{1-p}\lambda^{p/(p-1)}\left(\int_{S(f,s)\cap Q_{2r_i}(x_i)} f^p\,dx+4^m r_i^m s^p\right)\\
&\quad + s^{1-p}\lambda^{-p}\left(\int_{S(g,t)\cap Q_{2r_i}(x_i)} g^p\,dx+\frac{4^m r_i^m s^p}{c^p}\right)
\end{aligned}
$$

for any $\lambda > 0$. Thus

$$
\begin{aligned}
sr_i^m \le\ & (\theta^{1/p}+\lambda^{p/(p-1)})s^{1-p}\int_{S(f,s)\cap Q_{2r_i}(x_i)} f^p\,dx\\
&+a^{1/p}\int_{S(f,t)\cap Q_{2r_i}(x_i)} f\,dx+s^{1-p}\lambda^{-p}\int_{S(g,t)\cap Q_{2r_i}(x_i)} g^p\,dx\\
&+4^m\left(\theta^{1/p}+\lambda^{p/(p-1)}+\frac{a^{1/p}+\lambda^{-p}}{c^p}\right)r_i^m s.
\end{aligned}
$$

If $\theta \le 4^{-p(m+1)}$ and $\lambda \le 4^{-m-1}$, we can choose $c \ge 4^{-m-1}(a^{1/p}+\lambda^{-p})$ and it follows that

$$
\begin{aligned}
sr_i^m \le\ & 4(\theta^{1/p}+\lambda^{p/(p-1)})s^{1-p}\int_{S(f,s)\cap Q_{2r_i}(x_i)} f^p\,dx\\
&+4a^{1/p}\int_{S(f,t)\cap Q_{2r_i}(x_i)} f\,dx+4s^{1-p}\lambda^{-p}\int_{S(g,t)\cap Q_{2r_i}(x_i)} g^p\,dx. \quad (2.24)
\end{aligned}
$$

With the help of Vitali's covering lemma we can extract a subsequence $\{i_j\}$ (with j between 1 and $J \le \infty$) such that

$$
Q_{2r_{i_j}}(x_{i_j})\cap Q_{2r_{i_k}}(x_{i_k})=\emptyset \quad \text{for } j\ne k,
$$

but

$$
\sum_{i=1}^{I} r_i^m \le C\sum_{j=1}^{J} r_{i_j}^m
$$

for a constant C that depends only on m. Then it follows from (2.23) and (2.24) that

$$\begin{aligned}\int_{S(f,s)} f^p\,dx &\le Cs^p \sum_{j=1}^{J} r_{i_j}^m \\ &\le C(\theta^{1/p} + \lambda^{p/(p-1)}) \int_{S(f,s)} f^p\,dx + Cs^{p-1}a^{1/p} \int_{S(f,t)} f\,dx \\ &\quad + C\lambda^{-p} \int_{S(g,t)} g^p\,dx.\end{aligned}$$

If θ and λ are sufficiently small, then we conclude that

$$\int_{S(f,s)} f^p\,dx \le Cc^{p-1}a^{1/p}t^{p-1} \int_{S(f,t)} f\,dx + C\lambda^{-p} \int_{S(g,t)} g^p\,dx.$$

Clearly

$$\int_{S(f,t)} f^p\,dx \le \int_{S(f,s)} f^p\,dx + c^{p-1}t^{p-1} \int_{S(f,t)} f\,dx.$$

We therefore find that there exists a number C_1, depending only on m, p, and a, such that

$$\int_{S(f,t)} f^p\,dx \le C_1 \left(t^{p-1} \int_{S(f,t)} f\,dx + \int_{S(g,t)} g^p\,dx \right)$$

for all $t \ge t_0$, where

$$t_0 = \frac{1}{c} \left(\fint_{Q_1(0)} f^p\,dx \right)^{1/p}.$$

We can now apply Lemma 2.8 to the functions f and $g + t_0$. It follows that there exist two numbers $\bar{p} \in (p,q]$ and C_2, depending on m, p, q, and a, such that

$$\int_{Q_1(0)} f^{\bar{p}}\,dx \le C_2 \int_{Q_1(0)} (g^{\bar{p}} + t_0^{\bar{p}})\,dx.$$

The claim of the lemma follows immediately. □

In applications, we usually work with balls, not with cubes. For this reason we now restate Lemma 2.10, replacing the cubes with balls. A simple covering argument shows that the two versions are equivalent, thus there is no need to prove this one separately.

Lemma 2.11 *Suppose $1 < p < q$ and $a > 0$. There exist three numbers $\bar{p} \in (p, q]$, $\theta > 0$, and $b > 0$, depending only on m, p, q, and a, such that the following holds: Let $u \in L^p(B_1(0))$ and $v \in L^q(B_1(0))$ be non-negative functions satisfying*

$$\fint_{B_{r/2}(x_0)} u^p\,dx \le \theta \fint_{B_r(x_0)} u^p\,dx + a\left(\fint_{B_r(x_0)} u\,dx\right)^p + \fint_{B_r(x_0)} v^p\,dx$$

for every ball $B_r(x_0) \subset B_1(0)$. Then $u \in L^{\bar{p}}(B_{1/2}(0))$ and

$$\int_{B_{1/2}(0)} u^{\bar{p}}\,dx \le b\left(\int_{B_1(0)} u^p\,dx\right)^{\bar{p}/p} + b\int_{B_1(0)} v^{\bar{p}}\,dx.$$

Chapter 3

Harmonic Maps

Naturally, we first study the easiest of the problems mentioned in the introduction, the problem of harmonic maps. We begin this chapter by stating a few facts about embedded submanifolds of $\mathbb{R}^n$. In particular we examine the so-called nearest point projection and the second fundamental form associated to a submanifold, because we will need them to derive the basic properties of weakly harmonic maps. Then we give several variants of the Euler-Lagrange equation for the Dirichlet energy, and we derive a monotonicity formula for the energy of stationary weakly harmonic maps in concentric balls.

We begin the analysis of the regularity of weakly harmonic maps by proving that a continuous harmonic map is always smooth. Thus to prove regularity (or partial regularity) for a weakly harmonic map, it is sufficient to show that the map is continuous (away from a small set). We examine the question of continuity first in a special case. If the target manifold is the unit sphere $\mathbb{S}^{n-1}$, the arguments become considerably less involved and the underlying principles easier to understand. Therefore we first study this situation. After some further preparation, we can extend the arguments to the case of a general compact target manifold.

The case $N = \mathbb{S}^{n-1}$ also has the property that the notion of harmonic maps can be generalized naturally. We study this generalization in the last section of this chapter.

3.1 The Nearest Point Projection and the Second Fundamental Form

In order to understand weakly harmonic maps, we first have to know a few things about the embedded submanifold N in $\mathbb{R}^n$. We examine in particular the so-called nearest point projection onto N in this section and its relation to the second fundamental form. We will use these objects later in order to derive the Euler-Lagrange equation for critical points of the Dirichlet energy.

Since $N \subset \mathbb{R}^n$ is a smooth, compact submanifold, there exists a number $\kappa > 0$ such that in the κ-neighborhood

$$U_\kappa(N) = \{y \in \mathbb{R}^n : \ \operatorname{dist}(y, N) < \kappa\}$$

has the following property: For every $y \in U_\kappa(N)$ there exists a unique point $\pi_N(y) \in N$ such that

$$|y - \pi_n(y)| = \operatorname{dist}(y, N). \tag{3.1}$$

Definition 3.1 The map $\pi_N : U_\kappa(N) \to N$ defined by (3.1) is called the *nearest point projection* onto N.

If κ is chosen sufficiently small—which we assume in the following—then π_N has bounded derivatives of every order. We now examine the properties of the first and second derivatives.

Lemma 3.1 *For any $y \in N$, the first derivative $d\pi_N(y)$ is the orthogonal projection onto the tangent space T_yN.*

Proof. Suppose ν is a normal vector to N at y, i. e., $\nu \in (T_yN)^\perp$. Then $\pi_N(y + s\nu) = \pi_N(y)$ for any $s \in (-\kappa, \kappa)$. Thus $d\pi_N(y)\nu = 0$. If $X \in T_yN$, then

$$\lim_{s\to 0} \frac{1}{s} \operatorname{dist}(y + sX, N) = 0.$$

Hence

$$\begin{aligned}\lim_{s\to 0} \frac{1}{s}|\pi_N(y + sX) - \pi_N(y) - sX| &= \lim_{s\to 0} \frac{1}{s}|\pi_N(y + sX) - y - sX| \\ &= \lim_{s\to 0} \frac{1}{s} \operatorname{dist}(y + sX, N) = 0\end{aligned}$$

by (3.1). In other words, we have $d\pi_N(y)X = X$. □

Suppose now that y is a point of N and $X, Y \in T_yN$ two tangent vectors. Moreover, let $\nu \in (T_yN)^\perp$ be a normal vector at y. We can find smooth vector fields on $\mathbb{R}^n$ that agree with Y and ν, respectively, at y, such that the extension of Y is tangential and the extension of ν is normal everywhere on N. We use the same symbols, Y and ν, for these extensions.

The context will make clear in each case whether we mean the vector fields or their values at y.

We write D_X for the covariant derivative in $\mathbb{R}^n$ in the direction X; that is,

$$D_X Z = \sum_{i=1}^{n} X^i \frac{\partial Z}{\partial y^i}$$

for any differentiable vector field Z. By Lemma 3.1, we can represent the covariant derivative D_X^N on N in the direction X by the formula

$$D_X^N Y(y) = d\pi_N(y) D_X Y(y).$$

We also consider the expression

$$d\pi_N(y) D_X \nu(y). \tag{3.2}$$

If f is a smooth function on $\mathbb{R}^n$, we have

$$\begin{aligned} d\pi_N(y) D_X(f\nu)(y) &= f(y) d\pi_N(y) D_X \nu(y) + X(f)(y) d\pi_N(y)\nu(y) \\ &= f(y) d\pi_N(y) D_X \nu(y). \end{aligned}$$

That is, the expression (3.2) describes a tensor in ν. We conclude that there exists a unique section A of $T^*N \otimes T^*N \otimes (TN)^\perp$ such that

$$\langle \nu, A(y)(X,Y) \rangle = \langle D_X \nu, Y \rangle \tag{3.3}$$

for every $y \in N$ and all $X, Y \in T_y N$, $\nu \in (T_y N)^\perp$.

Definition 3.2 The tensor A defined by (3.3) is called the *second fundamental form* of the submanifold N in $\mathbb{R}^n$.

Equation (3.3) is one way to characterize the second fundamental form. We obtain another one by differentiating the identity

$$\langle \nu, Y \rangle = 0$$

in the direction of X. We see that

$$\langle D_X \nu, Y \rangle + \langle \nu, D_X Y \rangle = 0.$$

Thus we also have

$$\langle \nu, A(y)(X,Y) \rangle = - \langle D_X Y, \nu \rangle .$$

Moreover, we can express A in terms of the Hessian of the nearest point projection (which we denote by $\operatorname{Hess} \pi_N$).

Lemma 3.2 *The identities*

$$\operatorname{Hess} \pi_N(y)(X,Y) = -A(y)(X,Y)$$

and

$$\langle Y, \operatorname{Hess} \pi_N(y)(X,\nu)\rangle = -\langle \nu, A(y)(X,Y)\rangle$$

hold for all $y \in N$ *and* $X, Y \in T_yN$, $\nu \in (T_yN)^\perp$.

Proof. According to Lemma 3.1, we have

$$d\pi_N(y)\nu(y) = 0$$

at every point $y \in N$. We differentiate this identity in the direction X and obtain

$$\operatorname{Hess} \pi_N(y)(X,\nu) + d\pi_N(y)D_X\nu(y) = 0.$$

Taking the scalar product with Y already proves the second equation of the lemma. We can also differentiate the identity

$$d\pi_N(y)Y(y) = Y(y)$$

in the direction X and find that

$$\operatorname{Hess} \pi_N(y)(X,Y) + d\pi_N(y)D_XY(y) = D_XY(y).$$

It follows that $\operatorname{Hess} \pi_N(y)(X,Y)$ is in the normal space $(T_yN)^\perp$ and

$$\langle \nu, \operatorname{Hess} \pi_N(y)(X,Y)\rangle = -\langle \nu, A(y)(X,Y)\rangle .$$

The first equation also follows, and this concludes the proof. □

Remark 3.1 *The previous lemma also implies that the second fundamental form is symmetric, i. e.,*

$$A(y)(X,Y) = A(y)(Y,X)$$

for each pair of tangent vectors $X, Y \in T_yN$.

3.2 The Euler-Lagrange Equation

We now consider the energy functional

$$E(u) = \frac{1}{2} \int_\Omega |\nabla u|^2 \, dx$$

on the space $H^1(\Omega; N)$. For every $\phi \in C_0^\infty(\Omega; \mathbb{R}^n)$ and for every $s \in \mathbb{R}$ such that $|s|$ is sufficiently small, the map $u + s\phi$ takes on values in $U_\kappa(N)$. Thus

$$u_s = \pi_N \circ (u + s\pi)$$

is a well-defined map in $H^1(\Omega; N)$. We call u a critical point of E if

$$\left.\frac{d}{ds}\right|_{s=0} E(u_s) = 0 \tag{3.4}$$

for every such variation. An easy calculation yields

$$\left.\frac{d}{ds}\right|_{s=0} E(u_s) = \sum_{\alpha=1}^m \int_\Omega \left\langle \frac{\partial u}{\partial x^\alpha}, \frac{\partial \phi}{\partial x^\alpha} + \operatorname{Hess} \pi_N(u) \left(\phi, \frac{\partial u}{\partial x^\alpha} \right) \right\rangle dx.$$

We can split $\phi(x)$ into the tangential part

$$\phi^\top(x) = d\pi_N(u(x))\phi(x) \in T_{u(x)}N$$

and the normal part

$$\phi^\perp(x) = \phi(x) - \phi^\top(x) \in (T_{u(x)}N)^\perp$$

at almost every point $x \in \Omega$. Then we have

$$\left\langle \frac{\partial u}{\partial x^\alpha}, \operatorname{Hess} \pi_N(u) \left(\phi^\top, \frac{\partial u}{\partial x^\alpha} \right) \right\rangle = 0$$

almost everywhere by Lemma 3.2, because $\frac{\partial u}{\partial x^\alpha}$ and $\phi^\top$ are both tangent vectors. Using Lemma 3.2 again, we find

$$\begin{aligned} \left.\frac{d}{ds}\right|_{s=0} E(u_s) &= \sum_{\alpha=1}^m \int_\Omega \left\langle \frac{\partial u}{\partial x^\alpha}, \frac{\partial \phi}{\partial x^\alpha} + \operatorname{Hess} \pi_N(u) \left(\phi^\perp, \frac{\partial u}{\partial x^\alpha} \right) \right\rangle dx \\ &= \sum_{\alpha=1}^m \int_\Omega \left(\left\langle \frac{\partial u}{\partial x^\alpha}, \frac{\partial \phi}{\partial x^\alpha} \right\rangle - \left\langle \phi^\perp, A(u) \left(\frac{\partial u}{\partial x^\alpha}, \frac{\partial u}{\partial x^\alpha} \right) \right\rangle \right) dx. \end{aligned}$$

Because $A(u)(\frac{\partial u}{\partial x^\alpha}, \frac{\partial u}{\partial x^\alpha})$ is a normal vector almost everywhere, we can finally drop the superscript $\perp$ in the last term. To simplify the notation, we introduce the abbreviation

$$A(u)(\nabla u, \nabla u) = \sum_{\alpha=1}^{m} A(u)\left(\frac{\partial u}{\partial x^\alpha}, \frac{\partial u}{\partial x^\alpha}\right).$$

We have thus shown that condition (3.4) is equivalent to

$$\Delta u + A(u)(\nabla u, \nabla u) = 0 \quad \text{in } \Omega \tag{3.5}$$

in the weak sense.

Definition 3.3 A map $u \in H^1(\Omega; N)$ satisfying (3.5) is called *weakly harmonic.*

In general the expression

$$\tau(u) = \Delta u + A(u)(\nabla u, \nabla u)$$

is a distribution in Ω (with values in $\mathbb{R}^n$) that describes the negative L^2-gradient of the functional E at u. We have

$$\left.\frac{d}{ds}\right|_{s=0} E(\pi_N \circ (u + s\phi)) = -\int_\Omega \langle \tau(u), \phi \rangle \, dx$$

if $\tau(u) \in L^1_{\text{loc}}(\Omega; \mathbb{R}^n)$, and the corresponding formula with the right-hand side replaced by minus the value of the distribution $\tau(u)$ at ϕ, otherwise.

Definition 3.4 If $\tau(u) \in L^1_{\text{loc}}(\Omega; \mathbb{R}^n)$, then $\tau(u)$ is called the *tension field* of u.

There exists another condition which can also be used to characterize weakly harmonic maps. Obviously, (3.5) implies that Δu is represented by a function in $L^1(\Omega; \mathbb{R}^n)$ that satisfies

$$\Delta u \perp T_u N \quad \text{in } \Omega. \tag{3.6}$$

It turns out that the converse is also true.

Proposition 3.1 *The conditions (3.5) and (3.6) are equivalent.*

Proof. It suffices to show that (3.6) implies (3.5). To this end, consider the identity

$$d\pi_N(u)\nabla u = \nabla u.$$

Taking the divergence on both sides yields

$$\operatorname{Hess} \pi_N(u)(\nabla u, \nabla u) + d\pi_N(u)\Delta u = \Delta u.$$

If (3.6) is satisfied, however, then $d\pi_N(u)\Delta u = 0$, and it follows that

$$\Delta u = \operatorname{Hess} \pi_N(u)(\nabla u, \nabla u).$$

The first identity in Lemma 3.2 now implies (3.5). □

3.3 Stationary Weakly Harmonic Maps and the Monotonicity Formula

In the previous section, we defined weakly harmonic maps as critical points of E with respect to variations in the target manifold N. Instead, we can also consider variations of the domain Ω. For a given vector field $\psi \in C_0^\infty(\Omega; \mathbb{R}^m)$, we define

$$\Psi_s(x) = x + s\psi(x).$$

Whenever $|s|$ is sufficiently small, this is a diffeomorphism of Ω onto itself. Therefore, if $u \in H^1(\Omega; N)$, the maps

$$u_s = u \circ \Psi_s$$

are also in $H^1(\Omega; N)$. We now examine the condition

$$\left.\frac{d}{ds}\right|_{s=0} E(u_s) = 0 \tag{3.7}$$

for this kind of variation. We first compute

$$\nabla u_s(x) = \nabla u(\Psi_s(x)) + s \sum_{\alpha=1}^m \nabla \psi^\alpha(x) \frac{\partial u}{\partial x^\alpha}(\Psi_s(x)).$$

Hence

$$E(u_s) = \frac{1}{2} \int_\Omega |\nabla u \circ \Psi_s|^2 \, dx + s \sum_{\alpha,\beta=1}^m \int_\Omega \left\langle \frac{\partial u}{\partial x^\alpha} \circ \Psi_s, \frac{\partial u}{\partial x^\beta} \circ \Psi_s \right\rangle \frac{\partial \psi^\alpha}{\partial x^\beta} \, dx$$
$$+ \frac{s^2}{2} \sum_{\alpha,\beta=1}^m \int_\Omega \left\langle \frac{\partial u}{\partial x^\alpha} \circ \Psi_s, \frac{\partial u}{\partial x^\beta} \circ \Psi_s \right\rangle \nabla\psi^\alpha \cdot \nabla\psi^\beta \, dx.$$

To find the derivative with respect to s, we note that

$$\int_\Omega |\nabla u \circ \Psi_s|^2 \, dx = \int_\Omega |\nabla u|^2 |\det d\Psi_s^{-1}| \, dx.$$

Moreover, we have

$$\left.\frac{d}{ds}\right|_{s=0} |\det d\Psi_s^{-1}(x)| = -\operatorname{div}\psi(x).$$

Thus

$$\left.\frac{d}{ds}\right|_{s=0} E(u_s) = \sum_{\alpha,\beta=1}^{m} \int_\Omega \left\langle \frac{\partial u}{\partial x^\alpha}, \frac{\partial u}{\partial x^\beta} \right\rangle \frac{\partial \psi^\alpha}{\partial x^\beta}\, dx - \frac{1}{2}\int_\Omega |\nabla u|^2 \operatorname{div}\psi\, dx.$$

That is, the condition (3.7) is equivalent to

$$\operatorname{div}\left(|\nabla u|^2 e_\alpha - 2\left\langle \frac{\partial u}{\partial x^\alpha}, \nabla u \right\rangle \right) = 0 \quad \text{in } \Omega \tag{3.8}$$

for $\alpha = 1, \ldots, m$ in the distribution sense.

Definition 3.5 A weakly harmonic map $u \in H^1(\Omega; N)$ is called *stationary* if it satisfies (3.8).

If the map u is smooth, then (3.8) follows from the harmonic map equation (3.5). Indeed, for every smooth map $u : \Omega \to N$ we can simply calculate the divergence on the left-hand side of (3.8) and find

$$\operatorname{div}\left(|\nabla u|^2 e_\alpha - 2\left\langle \frac{\partial u}{\partial x^\alpha}, \nabla u \right\rangle \right) = -2\left\langle \frac{\partial u}{\partial x^\alpha}, \Delta u \right\rangle = -2\left\langle \frac{\partial u}{\partial x^\alpha}, \tau(u) \right\rangle,$$

because the other terms cancel and $\Delta u - \tau(u) \perp T_u N$. A map that is only in $H^1(\Omega; N)$, however, does not satisfy this equation in general, even if it is weakly harmonic. In other words: not every weakly harmonic map is stationary.

For the regularity theory, condition (3.8) is important above all because it implies a nice estimate for expressions of the form

$$r^{2-m} \int_{B_r(x_0)} |\nabla u|^2\, dx.$$

We usually refer to this as the rescaled energy of u in the ball $B_r(x_0)$. We will see that this quantity is non-decreasing in r if u is a stationary weakly harmonic map (or any map that satisfies (3.8)). This observation has already been made by R. Schoen and K. Uhlenbeck [1982; 1983] for minimizers of the Dirichlet energy. The version of the result we present here is due to P. Price [1983].

Lemma 3.3 *Let $u \in H^1(\Omega; N)$ be a map satisfying (3.8). Suppose $B_s(x_0) \subset B_r(x_0) \subset \Omega$. Then*

$$r^{2-m} \int_{B_r(x_0)} |\nabla u|^2 \, dx - s^{2-m} \int_{B_s(x_0)} |\nabla u|^2 \, dx = 2 \int_{B_r(x_0) \setminus B_s(x_0)} \frac{|(x - x_0) \cdot \nabla u|^2}{|x - x_0|^n} \, dx. \tag{3.9}$$

Proof. Formula (3.9) is invariant under translation, hence we may assume without loss of generality that $x_0 = 0$. We then test equation (3.8) with a function of the form $\eta(x) x^\alpha$, where $\eta \in C_0^\infty(\Omega)$. We obtain

$$\int_\Omega \left(x^\alpha \frac{\partial \eta}{\partial x^\alpha} |\nabla u|^2 - 2 x^\alpha \nabla \eta \cdot \left\langle \frac{\partial u}{\partial x^\alpha}, \nabla u \right\rangle \right) dx + \int_\Omega \eta \left(|\nabla u|^2 - 2 \left| \frac{\partial u}{\partial x^\alpha} \right|^2 \right) dx = 0$$

for $\alpha = 1, \ldots, m$. Summing over α yields

$$(m-2) \int_\Omega \eta |\nabla u|^2 \, dx + \int_\Omega x \cdot \nabla \eta \, |\nabla u|^2 \, dx = 2 \int_\Omega \langle x \cdot \nabla u, \nabla \eta \cdot \nabla u \rangle \, dx.$$

Suppose $B_\rho(0)$ is a ball contained in Ω. Let $\chi \in C^\infty(\mathbb{R}; [0,1])$ be a function satisfying $\chi \equiv 0$ in $(-\infty, 0]$ and $\chi \equiv 1$ in $[1, \infty)$. Set $\eta_k(x) = \chi(k(\rho - |x|))$. Now consider the last identity for $\eta = \eta_k$. Letting $k \to \infty$, we obtain in the limit

$$(m-2) \int_{B_\rho(0)} |\nabla u|^2 \, dx - \rho \int_{\partial B_\rho(0)} |\nabla u|^2 \, do = -\frac{2}{\rho} \int_{\partial B_\rho(0)} |x \cdot \nabla u|^2 \, do.$$

Thus

$$\frac{d}{d\rho} \left(\rho^{2-m} \int_{B_\rho(0)} |\nabla u|^2 \, dx \right) = 2 \int_{\partial B_\rho(0)} \frac{|x \cdot \nabla u|^2}{|x|^m} \, do.$$

Integrating this identity over the interval (s, r) gives (3.9). □

Remark 3.2 *Because the right-hand side of (3.9) is non-negative, we have in particular*

$$s^{2-m} \int_{B_s(x_0)} |\nabla u|^2 \, dx \le r^{2-m} \int_{B_r(x_0)} |\nabla u|^2 \, dx \tag{3.10}$$

for $0 < s \le r \le \operatorname{dist}(x_0, \partial\Omega)$ *(and equality holds if and only if* $(x-x_0)\cdot\nabla u = 0$ *in* $B_r(x_0)\backslash B_s(x_0)$*). For this reason, (3.9) is often called the monotonicity formula, and (3.10) is called the monotonicity inequality.*

Note also that inequality (3.10) is trivial if $m = 2$*. This is the reason why the case of a two-dimensional domain plays a special role in the regularity theory of harmonic maps. We will see that the monotonicity formula and condition (3.8) are not needed in order to prove regularity of harmonic maps if* $m = 2$*. We will say more about this point later.*

Lemma 3.4 *There exists a constant* C*, depending only on* m*, such that any solution* $u \in H^1(\Omega; N)$ *of (3.8) satisfies*

$$[u]_{\mathrm{BMO}(B_{r/2}(x_0))} \le C r^{2-m} \int_{B_r(x_0)} |\nabla u|^2 \, dx$$

for every ball $B_r(x_0) \subset \Omega$*.*

Proof. By Lemma 3.3,

$$\begin{aligned} s^{2-m} \int_{B_s(x_1)} |\nabla u|^2 \, dx &\le \left(\frac{r}{2}\right)^{2-m} \int_{B_{r/2}(x_1)} |\nabla u|^2 \, dx \\ &\le 2^{m-2} r^{2-m} \int_{B_r(x_0)} |\nabla u|^2 \, dx \end{aligned}$$

for every $x_1 \in B_{r/2}(x_0)$ and every $s \in (0, r/2]$. Thus the claim of the lemma follows from the Poincaré inequality. □

3.4 Higher Regularity

To prove regularity for a weakly harmonic map consists of two steps: to prove that it is continuous and to show that any continuous weakly harmonic map is smooth. The second step turns out to be the easier, therefore we discuss it first.

There are several ways to prove higher regularity for continuous weakly harmonic maps. The probably best-known proof uses arguments due to O. A. Ladyzhenskaya and N. N. Ural'tseva [1968] in order to estimate difference quotients of the gradient of u. This method is described, e. g., by J. Jost [2002] or by H.-J. Borchers and W.-D. Garber [1980] (for harmonic maps into spheres). We use a different method here. The arguments we present here are based on ideas of M. Feldman [1994]. (More precisely, they are the elliptic counterparts of arguments used by Feldman for the harmonic map heat flow.)

First we need some lemmas which give an estimate for the decay of the energy contained in concentric balls, as the radius tends to 0.

Lemma 3.5 *For every $\delta > 0$ and every $c_0 > 0$ there exist two constants $\epsilon > 0$ and $\theta \in (0,1)$, depending only on m, n, δ, and c_0, with the following property. Suppose $u \in H^1(B_r(x_0);\mathbb{R}^n)$ is a map with a weak Laplacian in $L^1(B_r(x_0);\mathbb{R}^n)$ satisfying*

$$|\Delta u| \leq c_0 |\nabla u|^2 \quad \textit{in } B_r(x_0)$$

and

$$r^{2-m} \int_{B_r(x_0)} |\nabla u|^2 \, dx \leq \epsilon.$$

Then

$$⨍_{B_{\theta r}(x_0)} |u - \bar{u}_{B_{\theta r}(x_0)}|^2 \, dx \leq \theta \delta r^{2-m} \int_{B_r(x_0)} |\nabla u|^2 \, dx.$$

Proof. We first note that all inequalities in the lemma are invariant under rescaling and translation. Thus we can assume without loss of generality that $r = 1$ and $x_0 = 0$.

We now argue by contradiction. Suppose there existed some numbers $\delta > 0$ and $c_0 > 0$ such that the claim were false. Then for any fixed $\theta \in (0, \frac{1}{2}]$ we could find a sequence of maps $u_i \in H^1(B_1(0);\mathbb{R}^n)$ satisfying

$$|\Delta u_i| \leq c_0 |\nabla u_i|^2 \quad \text{in } B_1(0)$$

and

$$\int_{B_1(0)} |\nabla u_i|^2 \, dx = \epsilon_i \to 0 \quad \text{as } i \to \infty,$$

but

$$⨍_{B_\theta(0)} |u_i - \bar{u}_i|^2 \, dx > \theta \delta \int_{B_1(0)} |\nabla u_i|^2 \, dx,$$

where

$$\bar{u}_i = ⨍_{B_\theta(0)} u_i \, dx.$$

Define

$$v_i = \frac{1}{\sqrt{\epsilon_i}}(u_i - \bar{u}_i).$$

Then obviously

$$\fint_{B_\theta(0)} v_i \, dx = 0.$$

Moreover, these maps satisfy

$$|\Delta v_i| \leq c_0 \sqrt{\epsilon_i} |\nabla v_i|^2 \quad \text{in } B_1(0) \tag{3.11}$$

and

$$\int_{B_1(0)} |\nabla v_i|^2 \, dx = 1,$$

but

$$\fint_{B_\theta(0)} |v_i|^2 \, dx > \theta\delta.$$

In particular the sequence $\{v_i\}_{i \in \mathbb{N}}$ is bounded in $H^1(B_1(0); \mathbb{R}^n)$. This implies that there exists a subsequence $\{v_{i_j}\}_{j \in \mathbb{N}}$ which converges weakly in $H^1(B_1(0); \mathbb{R}^n)$. Since $H^1(B_1(0); \mathbb{R}^n)$ is compactly embedded in the space $L^2(B_1(0); \mathbb{R}^n)$, we may assume that this subsequence also converges strongly in $L^2(B_1(0); \mathbb{R}^n)$. The limit map $v \in H^1(B_1(0); \mathbb{R}^n)$ has the properties

$$\fint_{B_\theta(0)} v \, dx = 0, \quad \|\nabla v\|_{L^2(B_1(0))} \leq 1,$$

and

$$\fint_{B_\theta(0)} |v|^2 \, dx \geq \theta\delta. \tag{3.12}$$

Moreover, inequality (3.11) and the fact that $\epsilon_i \to 0$ as $i \to \infty$ imply that

$$\begin{aligned}\left| \int_{B_1(0)} \nabla\phi \cdot \nabla v \, dx \right| &= \lim_{j\to\infty} \left| \int_{B_1(0)} \nabla\phi \cdot \nabla v_{i_j} \, dx \right| = \lim_{j\to\infty} \left| \int_{B_1(0)} \phi \, \Delta v_{i_j} \, dx \right| \\ &\leq c_0 \|\phi\|_{L^\infty(B_1(0))} \lim_{j\to\infty} \left(\sqrt{\epsilon_{i_j}} \int_{B_1(0)} |\nabla v_{i_j}|^2 \, dx \right) = 0\end{aligned}$$

for any $\phi \in C_0^\infty(B_1(0))$. That is, the map v satisfies Laplace's equation $\Delta v = 0$ in $B_1(0)$. Thus v is smooth, and its gradient satisfies Laplace's equation, too. Using the Poincaré inequality and the classical mean value inequality for solutions of Laplace's equation, we find that

$$\fint_{B_\theta(0)} |v|^2\,dx \le C\theta^{2-m}\int_{B_\theta(0)} |\nabla v|^2\,dx \le C\theta^2\int_{B_1(0)} |\nabla v|^2\,dx \le C\theta^2.$$

If $\theta < \delta/C$, we have a contradiction to (3.12), which finally proves the lemma. □

Lemma 3.6 *For every weakly harmonic map $u \in H^1(\Omega;N)\cap C^0(\Omega;N)$ and every $x_0 \in \Omega$ there exist two numbers $\gamma > 0$ and $r_0 \in (0, \frac{1}{2}\operatorname{dist}(x_0,\partial\Omega)]$ such that*

$$\sup_{x_1\in B_{r_0}(x_0)}\ \sup_{0<r\le r_0}\left(r^{1-m-\gamma}\int_{B_r(x_1)} |\nabla u|^2\,dx\right) < \infty. \tag{3.13}$$

Proof. Let $\lambda, \delta > 0$ be two small constants. Their values will be determined later. We choose r_0 such that $B_{3r_0}(x_0) \subset \Omega$ and

$$\operatorname*{osc}_{B_{3r_0}(x_0)} u \le \lambda. \tag{3.14}$$

Let $x_1 \in B_{r_0}(x_0)$ and $r \in (0, r_0]$. Choose a cut-off function $\xi \in C_0^\infty(B_{2r}(x_1))$ satisfying $0 \le \xi \le 1$ and $\xi \equiv 1$ in $B_r(x_1)$, such that $|\nabla\xi| \le 2/r$. Observing that $\nabla u = \nabla(u - \bar{u}_{B_{2r}(x_1)})$, we find

$$\begin{aligned}\int_{B_{2r}(x_1)} \xi^2|\nabla u|^2\,dx &= \int_{B_{2r}(x_1)} \xi^2\left\langle u - \bar{u}_{B_{2r}(x_1)}, A(u)(\nabla u,\nabla u)\right\rangle dx\\ &\quad - 2\int_{B_{2r}(x_1)} \xi\nabla\xi\cdot\left\langle u - \bar{u}_{B_{2r}(x_1)}, \nabla u\right\rangle dx\\ &\le C\lambda\left(\int_{B_{2r}(x_1)} \xi^2|\nabla u|^2\,dx + r^{m-2}\right).\end{aligned} \tag{3.15}$$

Here we have used first an integration by parts and then (3.14) and Young's inequality. If λ is sufficiently small, we conclude that

$$r^{2-m}\int_{B_r(x_1)} |\nabla u|^2\,dx \le C\lambda.$$

Suppose that we have furthermore $C\lambda \le \epsilon$ for the constant ϵ from Lemma 3.5 (belonging to the given δ). Then the lemma implies

$$\fint_{B_{\theta r}(x_1)} |u - \bar{u}_{B_{\theta r}(x_1)}|^2\,dx \le \theta\delta r^{2-m}\int_{B_r(x_1)} |\nabla u|^2\,dx$$

for some $\theta \in (0,1)$ that depends only on m, N, and δ.

We now choose another cut-off function $\eta \in C_0^\infty(B_{\theta r}(x_1))$ with $0 \le \eta \le 1$ and $\eta \equiv 1$ in $B_{\theta r/2}(x_1)$, such that $|\nabla \eta| \le 4/(\theta r)$. The same kind of estimate as in (3.15) and the inequality

$$-2 \int_{B_{\theta r}(x_1)} \eta \nabla \eta \cdot \langle u - \bar{u}_{B_{\theta r}(x_1)}, \nabla u \rangle \, dx \\ \le \frac{1}{4} \int_{B_{\theta r}(x_1)} \eta^2 |\nabla u|^2 \, dx + 4 \int_{B_{\theta r}(x_1)} |\nabla \eta|^2 |u - \bar{u}_{B_{\theta r}(x_1)}|^2 \, dx$$

yield

$$\int_{B_{\theta r}(x_1)} \eta^2 |\nabla u|^2 \, dx \le \left(C\lambda + \frac{1}{4} \right) \int_{B_{\theta r}(x_1)} \eta^2 |\nabla u|^2 \, dx \\ + \frac{C}{\theta^2 r^2} \int_{B_{\theta r}(x_1)} |u - \bar{u}_{B_{\theta r}(x_1)}|^2 \, dx.$$

If λ is small enough, it follows that

$$\theta^{1-m} \int_{B_{\theta r/2}(x_1)} |\nabla u|^2 \, dx \le C\delta \int_{B_r(x_1)} |\nabla u|^2 \, dx.$$

Choose δ so small that the factor $C\delta$ on the right-hand side is at most $\frac{1}{2}$. Then a repeated application of this inequality yields

$$\sup_{0<r\le r_0} \left(r^{1-m-\gamma} \int_{B_r(x)} |\nabla u|^2 \, dx \right) \le C_1$$

for certain constants C_1 and $\gamma > 0$ that may depend on u, x_0, and r_0, but not on x_1. This concludes the proof. □

Next, an estimate of the form (3.13) can be used to prove an L^∞-estimate for the gradient as follows.

Proposition 3.2 *Suppose a weakly harmonic map $u \in H^1(B_{2r_0}(x_0); N)$ satisfies (3.13) for some $\gamma > 0$. Then $\nabla u \in L^\infty(B_{r_0/2}(x_0); \mathbb{R}^{m\times n})$.*

Proof. First note that (3.13) and Hölder's inequality imply

$$\sup_{x\in B_{r_0}(x_0)} \sup_{0<r\le r_0} \left(r^{1/2-m-\gamma/2} \|\nabla u\|_{L^1(B_r(x))} \right) < \infty.$$

Moreover, the harmonic map equation (3.5) combined with (3.13) implies that

$$\sup_{x\in B_{r_0}(x_0)} \sup_{0<r\le r_0} \left(r^{1-m-\gamma}\|\Delta u\|_{L^1(B_r(x))}\right) < \infty.$$

Let now $\eta \in C_0^\infty(B_{r_0}(x_0))$ be a cut-off function satisfying $\eta \equiv 1$ in $B_{r_0/2}(x_0)$. Set $v = \eta u$. Then we have $\Delta v = \Delta\eta u + 2\nabla\eta \cdot \nabla u + \eta\Delta u$ and thus

$$\sup_{x\in \mathbb{R}^m} \sup_{0<r\le r_0} \left(r^{1-m-\gamma'}\|\Delta v\|_{L^1(B_r(x))}\right) = c_0 < \infty$$

for a certain number $c_0 > 0$ and $\gamma' = \min\{1, \gamma, (1+\gamma)/2\}$.

Suppose Γ is the fundamental solution of Laplace's equation, i. e.,

$$\Gamma(x) = \begin{cases} -\dfrac{1}{2\pi}\log|x| & \text{if } m = 2, \\ \dfrac{|x|^{2-m}}{m(m-2)\omega_m} & \text{if } m > 2. \end{cases}$$

Then for almost every $x \in B_{r_0}(x_0)$,

$$\begin{aligned} |\nabla v(x)| &= \left|\int_{\mathbb{R}^m} \nabla\Gamma(x-y)\Delta v(y)\,dy\right| \\ &\le \frac{1}{m\omega_m}\int_{\mathbb{R}^m} |x-y|^{1-m}|\Delta v(y)|\,dy \\ &= \frac{1}{m\omega_m}\int_0^{2r_0} r^{1-m}\int_{\partial B_r(x)} |\Delta v|\,do\,dr. \end{aligned}$$

An integration by parts then gives

$$\begin{aligned} |\nabla v(x)| \le{}& \frac{1}{m\omega_m}\lim_{r\searrow 0}\left(r^{1-m}\int_{B_r(x)}|\Delta v|\,dy\right) \\ &+ \frac{m-1}{m\omega_m}\int_0^{2r_0} r^{-m}\int_{B_r(x)}|\Delta v|\,dy\,dr \\ \le{}& \frac{c_0(m-1)}{m\omega_m}\int_0^{2r_0} r^{\gamma'-1}\,dr \le \frac{c_0(m-1)2^{\gamma'}r_0^{\gamma'}}{\gamma' m\omega_m}, \end{aligned}$$

and the proposition is proved. □

Theorem 3.1 *Let $u \in H^1(\Omega; N) \cap C^0(\Omega; N)$ be a weakly harmonic map. Then $u \in C^\infty(\Omega; N)$.*

Proof. Lemma 3.6 and Proposition 3.2 imply that $u \in W^{1,\infty}_{\mathrm{loc}}(\Omega; N)$. Using the harmonic map equation (3.5) and standard L^p-estimates for linear elliptic partial differential equations, we conclude that $u \in W^{2,p}_{\mathrm{loc}}(\Omega; N)$ for all $p < \infty$. By the Sobolev embedding theorem, $u \in C^1(\Omega; N)$. Differentiating (3.5) gives

$$|\Delta \nabla u| \leq C|\nabla u|(|\nabla u|^2 + |\nabla^2 u|).$$

We conclude that $u \in W^{3,p}_{\mathrm{loc}}(\Omega; N)$ for all $p < \infty$, and $u \in C^2(\Omega; N)$. By induction (differentiating (3.5) again with each step) we find that $u \in W^{l+1,p}_{\mathrm{loc}}(\Omega; N)$ for every $p < \infty$, and thus $u \in C^l(\Omega; N)$ for every $l \in \mathbb{N}$. □

3.5 The Target Manifold $\mathbb{S}^{n-1}$

Now we turn our attention to the question whether a stationary weakly harmonic map is continuous and thus smooth. We first consider stationary weakly harmonic maps in the special case that the target manifold is the standard sphere in $\mathbb{R}^n$, that is,

$$N = \mathbb{S}^{n-1} = \{y \in \mathbb{R}^n : \ |y| = 1\}.$$

The nearest point projection $\pi_{\mathbb{S}^{n-1}}$ for this manifold is

$$\pi_{\mathbb{S}^{n-1}}(y) = \frac{y}{|y|}, \quad y \in \mathbb{R}^n \backslash \{0\}.$$

We calculate

$$A(y)(X, Y) = -\operatorname{Hess} \pi_{\mathbb{S}^{n-1}}(y)(X, Y) = \langle X, Y \rangle \, y$$

for every $y \in \mathbb{S}^{n-1}$ and every pair of tangent vectors $X, Y \in T_y\mathbb{S}^{n-1}$. Thus the harmonic map equation in this situation is

$$\Delta u + |\nabla u|^2 u = 0 \quad \text{in } \Omega. \tag{3.16}$$

The special structure of the sphere, however, permits to write this in a different form. For this purpose, consider the exterior product $\wedge : \mathbb{R}^n \times \mathbb{R}^n \to \Lambda_2\mathbb{R}^n$. If u satisfies (3.16), then we compute

$$\operatorname{div}(u \wedge \nabla u) = 0 \quad \text{in } \Omega, \tag{3.17}$$

that is,

$$\operatorname{div}(u^i \nabla u^j - u^j \nabla u^i) = 0 \quad \text{in } \Omega$$

for $1 \le i, j \le m$, owing to the fact that $\langle u, \frac{\partial u}{\partial x^\alpha} \rangle = 0$ almost everywhere for each $\alpha = 1, \dots, m$. In fact, the two equations are equivalent.

Lemma 3.7 *A map $u \in H^1(\Omega; \mathbb{S}^{n-1})$ satisfies (3.16) if and only if it satisfies (3.17).*

Proof. For every $\phi \in C_0^\infty(\Omega; \mathbb{R}^n)$, we have

$$\begin{aligned}\int_\Omega |\nabla u|^2 \langle u, \phi \rangle \, dx &= \sum_{i,j=1}^n \int_\Omega (u^i \nabla u^j - u^j \nabla u^i) \cdot \nabla u^j \phi^i \, dx \\ &= \sum_{i=1}^n \int_\Omega \nabla u^i \cdot \nabla \phi^i \, dx \\ &\quad - \sum_{i,j=1}^n \int_\Omega \operatorname{div}(u^i \nabla u^j - u^j \nabla u^i) u^j \phi^i \, dx\end{aligned}$$

by the properties of the sphere and an integration by parts. The claim follows immediately. □

Equation (3.17) will be important for the proofs in this section. The arguments we use here are essentially due to F. Hélein [1990] (who proved regularity of weakly harmonic maps into spheres for $m = 2$) and L. C. Evans [1991] (who extended the results to other dimensions). We assume that $u \in H^1(\Omega; \mathbb{S}^{n-1})$ is a weakly stationary map, that is, it satisfies both (3.16) in the weak sense and (3.8) in the distribution sense. We consider balls $B_r(x_0)$ contained in Ω and study how the rescaled energy

$$r^{2-m} \int_{B_r(x_0)} |\nabla u|^2 \, dx$$

behaves as the radius r tends to 0. The aim is to show that if the rescaled energy around x_0 is small for a certain radius, then we obtain a decay that permits to apply Morrey's decay lemma (Lemma 2.1). Then we conclude that u is Hölder continuous. Higher regularity is implied by the results of Section 3.4.

Lemma 3.8 *There exists a constant C that depends only on m and n, such that every solution $u \in H^1(B_r(x_0); \mathbb{S}^{n-1})$ of (3.16) satisfies*

$$\begin{aligned}\left(\frac{r}{2}\right)^{2-m} \int_{B_{r/2}(x_0)} |\nabla u|^2 \, dx &\le \lambda r^{2-m} \int_{B_r(x_0)} |\nabla u|^2 \, dx \\ &\quad + C \fint_{B_r(x_0)} |u - \bar{u}_{B_r(x_0)}|^2 \, dx, \qquad (3.18)\end{aligned}$$

where

$$\lambda = C[u]_{\mathrm{BMO}(B_r(x_0))} + \frac{1}{4}.$$

Proof. We first note that all quantities in (3.18) remain invariant if we replace u by the map $u(rx+x_0)$ and $B_r(x_0)$ by $B_1(0)$. Thus we may assume without loss of generality that $x_0 = 0$ and $r = 1$.

Choose a cut-off function $\eta \in C_0^\infty(B_r(x_0))$ satisfying $\eta \equiv 1$ in $B_{1/2}(0)$, $0 \le \eta \le 1$ in $B_1(0)$, and $|\nabla\eta| \le 4$. Set

$$v = \eta(u - \bar{u}_{B_1(0)}).$$

Recall that

$$[v]_{\mathrm{BMO}(\mathbb{R}^m)} \le C_1[u]_{\mathrm{BMO}(B_1(0))} \tag{3.19}$$

for a constant C_1 that depends only on m, n, and the choice of η, by Lemma 2.6.

We use the fact that a weak solution of (3.16) satisfies (3.17), and an integration by parts, to derive the estimate

$$\begin{aligned}\int_{B_{1/2}(0)} |\nabla u|^2\,dx &\le \int_{B_1(0)} \eta|u \wedge \nabla u|^2\,dx \\ &= \int_{B_1(0)} \eta\,\langle u \wedge \nabla(u - \bar{u}_{B_1(0)}), u \wedge \nabla u\rangle\,dx \\ &= \int_{B_1(0)} \langle v \wedge \nabla u, u \wedge \nabla u\rangle\,dx \\ &\quad + \int_{B_1(0)} \nabla\eta \cdot \langle (u - \bar{u}_{B_1(0)}) \wedge u, u \wedge \nabla u\rangle\,dx. \end{aligned} \tag{3.20}$$

Here $\langle\,\cdot\,,\,\cdot\,\rangle$ denotes the scalar product in $\Lambda_2\mathbb{R}^n$ as well as the scalar product in $\mathbb{R}^m \otimes \Lambda_2\mathbb{R}^n$. We estimate the last term with Young's inequality:

$$\begin{aligned}&\int_{B_1(0)} \nabla\eta \cdot \langle (u - \bar{u}_{B_1(0)}) \wedge u, u \wedge \nabla u\rangle\,dx \\ &\quad\le \frac{1}{2^m}\int_{B_1(0)} |\nabla u|^2\,dx + 2^{m-2}\int_{B_1(0)} |\nabla\eta|^2|u - \bar{u}_{B_1(0)}|^2\,dx \\ &\quad\le \frac{1}{2^m}\int_{B_1(0)} |\nabla u|^2\,dx + 2^{m+2}\int_{B_1(0)} |u - \bar{u}_{B_1(0)}|^2\,dx. \end{aligned} \tag{3.21}$$

To estimate the first term on the right-hand side of (3.20), we first extend u to $\mathbb{R}^m$ such that

$$\int_{\mathbb{R}^m} |\nabla u|^2 \, dx \leq C \int_{B_1(0)} |\nabla u|^2 \, dx.$$

According to Theorem 2.3 and Theorem 2.4, we have

$$\int_{B_1(0)} \langle v \wedge \nabla u, u \wedge \nabla u \rangle \, dx \leq C[v]_{\mathrm{BMO}(\mathbb{R}^m)} \int_{B_1(0)} |\nabla u|^2 \, dx.$$

Using (3.19), we conclude that

$$\int_{B_1(0)} \langle v \wedge \nabla u, u \wedge \nabla u \rangle \, dx \leq C_2[u]_{\mathrm{BMO}(\mathbb{R}^m)} \int_{B_1(0)} |\nabla u|^2 \, dx \tag{3.22}$$

for a constant C_2 that depends only on m, n, and η. Inequality (3.18) now follows from (3.20)–(3.22). □

Lemma 3.9 *There exist $\epsilon_0 > 0$ and $\theta \in (0,1)$, both depending only on m and n, such that any stationary weakly harmonic map $u \in H^1(B_r(x_0); \mathbb{S}^{n-1})$ with*

$$r^{2-m} \int_{B_r(x_0)} |\nabla u|^2 \, dx \leq \epsilon_0$$

satisfies

$$(\theta r)^{2-m} \int_{B_{\theta r}(x_0)} |\nabla u|^2 \, dx \leq \frac{1}{2} r^{2-m} \int_{B_r(x_0)} |\nabla u|^2 \, dx. \tag{3.23}$$

Proof. We use inequality (3.18) from Lemma 3.8 for $B_{2\theta r}(x_0)$ instead of $B_r(x_0)$. The expression

$$⨍_{B_{2\theta r}(x_0)} |u - \bar{u}_{B_{2\theta r}(x_0)}|^2 \, dx$$

appearing in the last term can be estimated further with Lemma 3.5 (for 2θ instead of θ). If ϵ_0 and θ are sufficiently small, we conclude that

$$\begin{aligned}(\theta r)^{2-m} & \int_{B_{\theta r}(x_0)} |\nabla u|^2 \, dx \\ & \leq \left(C[u]_{\mathrm{BMO}(B_{2\theta r}(x_0))} + \frac{1}{4} \right) (2\theta r)^{2-m} \int_{B_{2\theta r}(x_0)} |\nabla u|^2 \, dx \\ & \quad + \frac{1}{8} r^{2-m} \int_{B_r(x_0)} |\nabla u|^2 \, dx.\end{aligned}$$

By the monotonicity formula (3.9),

$$(2\theta r)^{2-m} \int_{B_{2\theta r}(x_0)} |\nabla u|^2 \, dx \le r^{2-m} \int_{B_r(x_0)} |\nabla u|^2 \, dx.$$

Combined with the Poincaré inequality, the same formula implies

$$[u]_{\mathrm{BMO}(B_{2\theta r}(x_0))} \le C\sqrt{\epsilon_0}$$

if $\theta \le \frac{1}{4}$. If ϵ_0 is small enough, inequality (3.23) follows. □

Proposition 3.3 *There exists a number $\epsilon > 0$, depending only on m and n, such that any stationary weakly harmonic map $u \in H^1(B_r(x_0); \mathbb{S}^{n-1})$ satisfying*

$$r^{2-m} \int_{B_r(x_0)} |\nabla u|^2 \, dx \le \epsilon \tag{3.24}$$

is in the space $C^\infty(B_{r/2}(x_0); \mathbb{S}^{n-1})$.

Proof. If we choose $\epsilon = 2^{2-m}\epsilon_0$ for the constant ϵ_0 from Lemma 3.9, then we can apply the lemma for every ball $B_{r/2}(x_1)$ with $x_1 \in B_{r/2}(x_0)$ and $0 < s \le r/2$. We obtain

$$(\theta r)^{2-m} \int_{B_{\theta r}(x_0)} |\nabla u|^2 \, dx \le \frac{1}{2}\epsilon_0.$$

In particular Lemma 3.9 can now be applied to the ball $B_{\theta r}(x_1)$. Proceeding inductively, we see that

$$(\theta^k r)^{2-m} \int_{B_{\theta^k r}(x_0)} |\nabla u|^2 \, dx \le 2^{-k}\epsilon_0.$$

Hence there exists a number $\alpha > 0$ such that

$$\sup_{x_1 \in B_{r/2}(x_0)} \sup_{0 < s \le r/2} \left(s^{2-m-\alpha} \int_{B_s(x_1)} |\nabla u|^2 \, dx \right) < \infty.$$

By Lemma 2.1, the map u is Hölder continuous in $B_{r/2}(x_0)$. Higher regularity then follows from Theorem 3.1. □

We want to use Proposition 3.3 in order to say something about the regularity of stationary weakly harmonic maps even if the energy is not small. Then of course the condition (3.24) is not true for every ball in the domain Ω. The set of points $x_0 \in \Omega$, however, such that it is true for no ball around x_0, is small. We have the following result.

Theorem 3.2 *Suppose $u \in H^1(\Omega;\mathbb{S}^{n-1})$ is a stationary weakly harmonic map. There exists a relatively closed set $\Sigma \subset \Omega$ of vanishing $(m-2)$-dimensional Hausdorff measure, such that $u \in C^\infty(\Omega\backslash\Sigma;\mathbb{S}^{n-1})$.*

Proof. In view of Proposition 3.3, the natural candidate for Σ is the set

$$\Sigma = \{x_0 \in \Omega : \ \Theta(x_0) \geq \epsilon\},$$

where ϵ is the number from Proposition 3.3 and

$$\Theta(x_0) = \lim_{r \searrow 0}\left(r^{2-m}\int_{B_r(x_0)} |\nabla u|^2\,dx\right).$$

The monotonicity formula not only implies that Θ is well-defined everywhere in Ω, but also that it is upper semi-continuous. Indeed, if we have a sequence of points $x_i \in \Omega$ that converge to $x_0 \in \Omega$, then for any $\delta > 0$ we can choose a radius $r > 0$ such that $B_r(x_0) \subset \Omega$ and

$$r^{2-m}\int_{B_r(x_0)} |\nabla u|^2\,dx \leq \Theta(x_0) + \delta.$$

Then for any sufficiently large i, we have $x_i \in B_r(x_0)$. Hence

$$(r-|x_i-x_0|)^{2-m}\int_{B_{r-|x_i-x_0|}(x_i)} |\nabla u|^2\,dx \leq \left(1-\frac{|x_i-x_0|}{r}\right)^{2-m}(\Theta(x_0)+\delta).$$

The monotonicity formula (3.9) implies

$$\Theta(x_i) \leq \left(1-\frac{|x_i-x_0|}{r}\right)^{2-m}(\Theta(x_0)+\delta),$$

and because δ was chosen arbitrarily, it follows that

$$\limsup_{i\to\infty}\Theta(x_i) \leq \Theta(x_0).$$

Hence Σ is a relatively closed set. By Proposition 3.3, we have $u \in C^\infty(\Omega\backslash\Sigma;\mathbb{S}^{n-1})$. It only remains to prove that the $(m-2)$-dimensional Hausdorff measure of Σ is 0.

To this end, it suffices to prove the same statement for the intersection of Σ with an arbitrary compact subset $\Sigma' \subset \Sigma$, for Σ is a countable union of such sets. Let thus $\Sigma' \subset \Sigma$ be compact. For a given $R > 0$, consider coverings of Σ' with countable collections of balls $B_{r_i}(x_i)$, $i \in \mathbb{N}$, such that $x_i \in \Sigma'$ and $0 < r_i \leq R$. By Vitali's covering lemma, any such collection has a finite subcollection (we may assume it consists of the balls $B_{r_i}(x_1), \dots, B_{r_I}(x_I)$ for some $I \in \mathbb{N}$) such that

$$B_{r_i}(x_i) \cap B_{r_j}(x_j) = \emptyset \quad \text{if } 1 \le i, j \le I,\ i \ne j,$$

and

$$\Sigma' \subset \bigcup_{i=1}^{I} B_{5r_i}(x_i).$$

Since each x_i is in Σ, the monotonicity formula implies

$$r_i^{2-m} \int_{B_{r_i}(x_0)} |\nabla u|^2 \, dx \ge \epsilon.$$

Hence

$$\sum_{i=1}^{I} (5r_i)^{m-2} \le \frac{5^{m-2}}{\epsilon} \sum_{i=1}^{I} \int_{B_{r_i}(x_i)} |\nabla u|^2 \, dx \le \frac{5^{m-2}}{\epsilon} \int_{\Omega} |\nabla u|^2 \, dx.$$

Letting $R \searrow 0$, we conclude that the $(m-2)$-dimensional Hausdorff measure of Σ' is finite.

For $R > 0$, define now

$$V_R(\Sigma') = \{x \in \Omega : \ \operatorname{dist}(x, \Sigma') < R\}.$$

Then with the same arguments as above, we see that the $(m-2)$-dimensional Hausdorff measure of Σ' is at most a constant times the number

$$\lim_{R \searrow 0} \int_{V_R(\Sigma')} |\nabla u|^2 \, dx.$$

Because the Hausdorff dimension of Σ' is at most $m-2$, the last expression must be 0. This completes the proof. □

Finally we give a little special attention to the case $m = 2$. As already pointed out, the monotonicity inequality (3.9), which follows from condition (3.8) for stationary weakly harmonic maps in higher dimensions, is trivial in two dimensions. But this inequality is the only reason we consider stationary weakly harmonic maps at all. The condition is used in no other way in the proof of Theorem 3.2. Hence for $m = 2$, there is no need to impose (3.8). Moreover, if the 0-dimensional Hausdorff measure of a set vanishes, it follows that the set is empty. Thus the arguments of this section also prove the following theorem.

Theorem 3.3 *Suppose $m = 2$. Then any weakly harmonic map $u \in H^1(\Omega; \mathbb{S}^{n-1})$ is in $C^\infty(\Omega; \mathbb{S}^{n-1})$.*

3.6 A Moving Tangent Frame Field

We now consider other target manifolds, that is, N is henceforth again a general compact, smooth Riemannian manifold without boundary, isometrically embedded in $\mathbb{R}^n$. The aim of this and the following sections is to prove a partial regularity result for stationary weakly harmonic maps in this situation. It was crucial for the proofs in the last section that the harmonic map equation (3.16) for maps into a sphere can be rewritten in the form (3.17). Such a reformulation of the equation is no longer available for more general target manifolds, but the following construction provides a substitute for it. It is based on ideas and arguments of F. Hélein [1991a].

Suppose $u \in H^1(\Omega; N)$ is a stationary weakly harmonic map, that is, a solution of

$$\Delta u \perp T_u N \quad \text{in } \Omega \tag{3.25}$$

satisfying

$$\operatorname{div}\left(|\nabla u|^2 e_\alpha - 2\left\langle \frac{\partial u}{\partial x^\alpha}, \nabla u \right\rangle\right) = 0 \quad \text{in } \Omega \tag{3.26}$$

for $\alpha = 1, \ldots, m$. We denote the dimension of N by k. We want to construct a tangent frame field $(e_1, \ldots, e_k)$ along u, that is, a collection of k maps $e_1, \ldots, e_k \in H^1(\Omega; \mathbb{S}^{n-1})$ such that for almost every $x \in \Omega$, the vectors $e_1(x), \ldots, e_k(x)$ constitute an orthonormal basis of $T_{u(x)}N$. This is no problem if N possesses smooth vector fields $\bar{e}_1, \ldots, \bar{e}_k : N \to \mathbb{S}^{n-1}$ which give an orthonormal basis of the tangent space at each point of N. In this case we can simply set $e_i = \bar{e}_i \circ u$; or, more generally,

$$e_i = \sum_{j=1}^{k} R_{ij}(\bar{e}_j \circ u)$$

for any $R = (R_{ij}) \in H^1(\Omega; \mathrm{SO}(k))$. If N does not have this property, the situation is a little more complicated.

The basic idea in the latter case is to replace N by another manifold that is easier to work with. More precisely, we construct a smooth embedding $\Phi : N \to \tilde{N}$ for some compact, smooth Riemannian manifold $\tilde{N}$, such that the image of N is contained in one single coordinate chart of $\tilde{N}$. Then it is easy to construct an orthonormal tangent frame field $(\tilde{e}_1, \ldots, \tilde{e}_{\tilde{k}})$ (where $\tilde{k}$ is the dimension of $\tilde{N}$) along the map $\tilde{u} = \Phi \circ u$ into the manifold $\tilde{N}$, for instance by a Gram-Schmidt orthogonalization process. Of course this does not mean that the same holds for the map u and the manifold N. But if we can construct Φ such that $\tilde{u}$ is again a stationary weakly harmonic map, then we can work with $\tilde{u}$ thereafter, instead of u. After all, if we can prove partial regularity for $\tilde{u}$, then the same kind of regularity follows for u. In other words, if this construction is successful, we may assume without loss of generality that N has the property mentioned above.

The manifold $\tilde{N}$ and the embedding Φ can be constructed as follows: First choose a tubular neighborhood

$$U = U_\kappa(N) = \{y \in \mathbb{R}^n : \ \operatorname{dist}(y, N) < \kappa\}$$

such that the nearest point projection $\pi_N : U \to N$ is smooth on U with bounded derivatives of any order. On $U \backslash N$, the map

$$U \backslash N \to N \times (0, \kappa), \quad y \mapsto (\pi_N(y), |y - \pi_N(y)|)$$

is a local diffeomorphism. It induces a Riemannian metric on $U \backslash N$, and this metric can be extended smoothly throughout U. The set U together with this metric gives a Riemannian manifold.

This manifold can be extended to a compact, smooth Riemannian manifold $\tilde{N}$, for instance by gluing another copy of the same manifold to its boundary (or by regarding U as a piece of an n-torus and extending the metric smoothly). The manifold $\tilde{N}$ can be embedded in a Euclidean space $\mathbb{R}^{\tilde{n}}$ for some positive integer $\tilde{n}$ by the Nash–Moser embedding theorem (see J. Nash [1956]). It is clear that this manifold $\tilde{N}$ and the embedding $\Phi : N \to \tilde{N}$ given by the inclusion $N \subset U$ have all the desired properties, as soon as we have verified that $\tilde{u} = \Phi \circ u$ is a stationary weakly harmonic map. We only have to check whether it is a weakly harmonic map, however, since the condition (3.26) is equivalent to (3.7). Passing from u to $\tilde{u}$ does not change anything in (3.7), since only variations of the domain are involved there.

To see that $\tilde{u}$ is weakly harmonic, we observe that N is mapped onto a totally geodesic submanifold of $\tilde{N}$ by the embedding Φ (i. e., every geodesic in N is mapped onto a geodesic in $\tilde{N}$) by the construction of $\tilde{N}$. This fact can be expressed by the condition

$$\operatorname{Hess}\Phi(y)(X, Y) \perp T_{\Phi(y)}\tilde{N}$$

for any $y \in N$ and any pair of tangent vectors $X, Y \in T_yN$. (Here we regard Φ as a map into $\mathbb{R}^{\tilde{n}}$.) By the properties of the nearest point projection, the same is true for the extension of Φ to U given by the composition with π_N. We denote this extension again by Φ. Thus

$$\Delta\tilde{u} = d\Phi(u)\Delta u + \sum_{\alpha=1}^{m} \operatorname{Hess}\Phi(u)\left(\frac{\partial u}{\partial x^\alpha}, \frac{\partial u}{\partial x^\alpha}\right) \perp T_{\tilde{u}}\tilde{N}.$$

This finishes the construction of $\tilde{N}$ and Φ. We now drop the tildes and write u instead of $\tilde{u}$, N instead of $\tilde{N}$, k instead of $\tilde{k}$, and $\mathbb{R}^n$ instead of $\mathbb{R}^{\tilde{n}}$ in the rest of this section and in the following sections.

Remark 3.3 *We can do the same construction for a map $u \in H^1(\Omega; N)$ with a non-vanishing tension field $\tau(u)$. In this case, the computations give $\tilde{\tau}(\tilde{u}) = d\Phi(u)\tau(u)$, where $\tilde{\tau}(\tilde{u})$ is the tension field of $\tilde{u}$ with respect to the manifold $\tilde{N}$. We will make use of this fact in the next chapter.*

Now we can assume that we have an orthonormal tangent frame field $(\bar{e}_1 \circ u, \ldots, \bar{e}_k \circ u)$ along u. For any $R \in H^1(\Omega; \mathrm{SO}(k))$, the maps

$$e_i = \sum_{j=1}^{k} R_{ij}(\bar{e}_j \circ u) \tag{3.27}$$

give another orthonormal tangent frame field. Suppose a ball $B_r(x_0) \subset \Omega$ is given. Then we can choose $R \in H^1(B_r(x_0); \mathrm{SO}(k))$ such that the functional

$$I(R) = \frac{1}{2}\sum_{i=1}^{k} \int_{B_r(x_0)} |\nabla e_i|^2 \, dx \tag{3.28}$$

is minimized, where e_i is given by (3.27). Such a map can be constructed, e. g., by the direct method in the calculus of variations: It is readily checked that

$$\int_{B_r(x_0)} |\nabla R|^2 \, dx \le CI(R) + C\int_{B_r(x_0)} |\nabla u|^2 \, dx.$$

Hence if $\{R_l\}_{l\in\mathbb{N}} \subset H^1(B_r(x_0); \mathrm{SO}(k))$ is a minimizing sequence for I, then this sequence is bounded in $H^1(B_r(x_0); \mathbb{R}^{k\times k})$. A subsequence is weakly convergent in $H^1(B_r(x_0); \mathbb{R}^{k\times k})$. We may also assume that we have convergence pointwise almost everywhere, therefore the limit R is in $H^1(B_r(x_0); \mathrm{SO}(k))$. This map R is obviously a minimizer of I.

Next we want to find the Euler-Lagrange equations for this minimizer. To this end, consider the Lie algebra $\mathfrak{so}(k)$ belonging to the Lie group

$\mathrm{SO}(k)$, i. e., $\mathfrak{so}(k)$ is the space of all skew symmetric $(k \times k)$-matrices. We consider variations of R of the form

$$R^s = \exp(sT)R, \quad s \in \mathbb{R},$$

where $T \in C^\infty(\mathbb{R}^m; \mathfrak{so}(k))$. We have

$$\left.\frac{d}{ds}\right|_{s=0} \nabla \exp(sT) = \nabla T$$

and thus

$$\left.\frac{d}{ds}\right|_{s=0} \nabla R^s = \nabla T\, R + T\, \nabla R.$$

If we define

$$e_i^s = \sum_{j=1}^{k} R_{ij}^s(\bar{e}_j \circ u),$$

it follows that

$$\left.\frac{d}{ds}\right|_{s=0} \nabla e_i^s = \sum_{j=1}^{k} \nabla T_{ij} e_j + T_{ij} \nabla e_j.$$

Hence

$$0 = \left.\frac{d}{ds}\right|_{s=0} I(R^s) = \sum_{i,j=1}^{k} \int_{B_r(x_0)} (\nabla T_{ij} \cdot \langle e_j, \nabla e_i \rangle + T_{ij} \langle \nabla e_j, \nabla e_i \rangle)\, dx.$$

Now we use the facts that T is skew symmetric, and that

$$\langle e_i, \nabla e_j \rangle = - \langle \nabla e_i, e_j \rangle.$$

We conclude that

$$0 = \sum_{i<j} \int_{B_r(x_0)} \nabla T_{ij} \cdot \langle \nabla e_i, e_j \rangle \, dx$$

for all $T \in C^\infty(\mathbb{R}^m; \mathfrak{so}(k))$. Thus we have the equations

$$\operatorname{div} \langle \nabla e_i, e_j \rangle = 0 \quad \text{in } B_r(x_0)$$

and the boundary conditions

$$(x - x_0) \cdot \langle \nabla e_i, e_j \rangle = 0 \quad \text{on } \partial B_r(x_0).$$

Defining

$$\omega_{ij}(x) = \begin{cases} \langle \nabla e_i(x), e_j(x) \rangle & \text{if } x \in B_r(x_0), \\ 0 & \text{if } x \notin B_r(x_0), \end{cases} \tag{3.29}$$

we can write this in the form

$$\operatorname{div} \omega_{ij} = 0 \quad \text{in } \mathbb{R}^m. \tag{3.30}$$

This equation is to be understood in the distribution sense.

If u is weakly harmonic, i. e., if (3.25) is satisfied, then we have furthermore

$$\operatorname{div} \langle \nabla u, e_i \rangle = \langle \Delta u, e_i \rangle + \langle \nabla u, \nabla e_i \rangle = \sum_{j=1}^{k} \langle \nabla u, e_j \rangle \cdot \omega_{ij}. \tag{3.31}$$

This equation will eventually play the role of (3.17) when we want to prove partial regularity results for stationary weakly harmonic maps into general target manifolds.

3.7 General Target Manifolds

The tangent frame field $(e_1, \ldots, e_k)$ constructed in the previous chapter was used by F. Hélein [1991a] to prove regularity of weakly harmonic maps on two-dimensional domains. Using the same ideas (and a few additional arguments), F. Bethuel [1993] proved partial regularity for stationary weakly harmonic maps into general compact target manifolds in higher dimensions. The arguments we present in this section are essentially the same as used by F. Bethuel.

We first recall the situation we consider. We assume that N is a smooth, compact Riemannian manifold without boundary of dimension k and $u \in H^1(\Omega; N)$ a stationary weakly harmonic map. That is, it satisfies

$$\Delta u + A(u)(\nabla u, \nabla u) = 0 \quad \text{in } \Omega$$

and

$$\operatorname{div}\left(|\nabla u|^2 e_\alpha - 2 \left\langle \frac{\partial u}{\partial x^\alpha}, \nabla u \right\rangle \right) = 0 \quad \text{in } \Omega \tag{3.32}$$

for $\alpha = 1, \ldots, n$. By the construction of Section 3.6, we may assume without loss of generality that a smooth orthonormal tangent frame field $(\bar{e}_1, \ldots, \bar{e}_k)$ exists at least in some neighborhood of the closure of $u(\Omega)$. Hence for any ball $B_r(x_0) \subset \Omega$, we can choose a minimizer $R \in H^1(B_r(x_0); \mathrm{SO}(k))$ of the functional defined in (3.28) and use the properties of the vector fields ω_{ij} given by (3.29).

The general strategy is now the same as in Section 3.5: Assuming that we have small scaled energy in some ball, we try to find a decay estimate that permits to apply Morrey's decay lemma, so that we obtain Hölder continuity of u. This time, however, we work with the L^1-norm of the gradient of u in order to avoid certain technical problems. The next lemma is the first step towards this end.

Lemma 3.10 *There exists a constant C, depending only on m and N, such that the following holds. Suppose $u \in H^1(B_r(x_0); N)$ is a weakly harmonic map with*

$$r^{2-m} \int_{B_r(x_0)} |\nabla u|^2 \, dx \le \epsilon.$$

Then

$$(\theta r)^{1-m} \int_{B_{\theta r}(x_0)} |\nabla u| \, dx \le C(\sqrt{\epsilon} + \epsilon)[u]_{\mathrm{BMO}(B_r(x_0))} + C\theta r^{1-m} \int_{B_r(x_0)} |\nabla u| \, dx \tag{3.33}$$

for any $\theta \in (0,1)$.

Proof. Let $(e_1, \ldots, e_k)$ be the orthonormal tangent frame field constructed in Section 3.6. That is, the vectors $e_1(x), \ldots, e_k(x)$ form an orthonormal basis of $T_{u(x)}N$ at almost every $x \in \Omega$, and the vector fields

$$\omega_{ij} = \begin{cases} \langle \nabla e_i, e_j \rangle & \text{in } B_r(x_0), \\ 0 & \text{in } \mathbb{R}^m \backslash B_r(x_0), \end{cases}$$

satisfy

$$\operatorname{div} \omega_{ij} = 0 \quad \text{in } \mathbb{R}^m.$$

Moreover, we have

$$\sum_{i=1}^{k} \int_{B_r(x_0)} |\nabla e_i|^2 \, dx \le C \int_{B_r(x_0)} |\nabla u|^2 \, dx. \tag{3.34}$$

We choose a cut-off function $\eta \in C_0^\infty(B_r(x_0))$ satisfying $\eta \equiv 1$ in $B_{r/2}(x_0)$ and $0 \le \eta \le 1$ in $B_r(x_0)$, such that $|\nabla \eta| \le 4/r$. We apply the Hodge decomposition of Theorem 2.1 to the vector fields

$$\alpha_i = \left\langle \nabla(\eta(u - \bar{u}_{B_r(x_0)})), e_i \right\rangle, \quad i = 1, \ldots, k.$$

The theorem implies that there exist $a_i \in H^1(\mathbb{R}^m)$ and $A_i \in L^2(\mathbb{R}^m; \mathbb{R}^m)$ satisfying

$$\alpha_i = \nabla a_i + A_i, \quad i = 1, \ldots, k, \tag{3.35}$$

such that $\operatorname{div} A_i = 0$ in $\mathbb{R}^m$ and

$$\int_{\mathbb{R}^m} (|\nabla a_i|^2 + |A_i|^2)\, dx \leq C \int_{B_r(x_0)} |\nabla u|^2\, dx. \tag{3.36}$$

In $B_{r/2}(x_0)$, each α_i coincides with $\langle \nabla u, e_i \rangle$, thus we have

$$|\nabla u|^2 = \sum_{i=1}^{k} |\alpha_i|^2 \quad \text{in } B_{r/2}(x_0).$$

Moreover,

$$\operatorname{div} \alpha_i = \sum_{j=1}^{k} \alpha_j \cdot \omega_{ij} \quad \text{in } B_{r/2}(x_0)$$

by (3.31). It follows that

$$\Delta a_i = \sum_{j=1}^{k} \alpha_j \cdot \omega_{ij} \quad \text{in } B_{r/2}(x_0). \tag{3.37}$$

Choose another cut-off function $\xi \in C_0^\infty(B_{r/2}(x_0))$ with $\xi \equiv 1$ in $B_{r/4}(x_0)$ and $0 \leq \xi \leq 1$, such that $|\nabla \xi| \leq 8/r$. Then

$$\int_{B_{r/4}(x_0)} |\nabla u|\, dx \leq \sum_{i=1}^{k} \int_{\mathbb{R}^m} \xi |\alpha_i|\, dx \leq \sum_{i=1}^{k} \int_{\mathbb{R}^m} \xi(|\nabla a_i| + |A_i|)\, dx. \tag{3.38}$$

We estimate the integrals on the right-hand side further, first the ones with the integrands $\xi|A_i|$. We have

$$\int_{\mathbb{R}^m} \xi |A_i|\, dx \leq (r/2)^{m/2} \sqrt{\omega_m} \left(\int_{\mathbb{R}^m} |A_i|^2\, dx \right)^{1/2} \tag{3.39}$$

and

$$\int_{\mathbb{R}^m} |A_i|^2\, dx = \int_{B_r(x_0)} A_i \cdot \langle \nabla(\eta(u - \bar{u}_{B_r(x_0)})), e_i \rangle\, dx. \tag{3.40}$$

To see that (3.40) is true, let $\zeta_R \in C_0^\infty(B_{2R}(0))$ be a cut-off function with $\zeta_R \equiv 1$ in $B_R(0)$ and $|\nabla\zeta| \leq 2/R$. Then we find

$$\begin{aligned}\left|\int_{\mathbb{R}^m} A_i \cdot \nabla a_i \, dx\right| &= \lim_{R\to\infty}\left|\int_{\mathbb{R}^m} \zeta_R A_i \cdot \nabla a_i \, dx\right| \\ &= \lim_{R\to\infty}\left|\int_{\mathbb{R}^m} \nabla\zeta_R \cdot A_i \left(a_i - \overline{(a_i)}_{B_R(0)}\right) dx\right| \\ &\leq \liminf_{R\to\infty}\left(\int_{\mathbb{R}^m\setminus B_R(0)} |A_i|^2 \, dx\right)^{1/2}\left(\int_{\mathbb{R}^m} |\nabla a_i|^2 \, dx\right)^{1/2},\end{aligned}$$

using the Hölder inequality and the Poincaré inequality in the last step. The last expression is 0, because $A_i, \nabla a_i \in L^2(\mathbb{R}^m;\mathbb{R}^m)$. Thus (3.40) follows from (3.35).

An integration by parts on the right-hand side of (3.40) yields

$$\int_{\mathbb{R}^m} |A_i|^2 \, dx = -\int_{\mathbb{R}^m} A_i \cdot \langle \eta(u - \bar{u}_{B_r(x_0)}), \nabla e_i \rangle \, dx. \tag{3.41}$$

We extend e_i to $\mathbb{R}^m$ such that

$$\|\nabla e_i\|^2_{L^2(\mathbb{R}^m)} \leq C \int_{B_r(x_0)} |\nabla u|^2 \, dx. \tag{3.42}$$

This is possible by (3.34). Then the right-hand side of (3.41) is of the right form to apply Theorem 2.3 and Theorem 2.4. We obtain the estimate

$$\int_{\mathbb{R}^m} |A_i|^2 \, dx \leq C[\eta(u - \bar{u}_{B_r(x_0)})]_{\mathrm{BMO}(\mathbb{R}^m)} \|\nabla e_i\|_{L^2(\mathbb{R}^m)} \left(\int_{\mathbb{R}^m} |A_i|^2 \, dx\right)^{1/2}.$$

We use (3.42) and Lemma 2.6 to estimate two of the factors on the right-hand side. Using (3.39), we then conclude that

$$\int_{\mathbb{R}^m} \xi |A_i| \, dx \leq C[u]_{\mathrm{BMO}(B_r(x_0))} \left(r^m \int_{B_r(x_0)} |\nabla u|^2 \, dx\right)^{1/2}. \tag{3.43}$$

Next we estimate the other part on the right-hand side of (3.38). To this end, we consider the solutions b_i of the boundary value problems

$$\begin{cases} \Delta b_i = 0 & \text{in } B_{r/2}(x_0), \\ b_i = a_i & \text{on } \partial B_{r/2}(x_0). \end{cases} \tag{3.44}$$

Furthermore, let ψ_i be the solutions of

$$\begin{cases} \Delta\psi_i = -\operatorname{div}\left(\dfrac{\nabla a_i - \nabla b_i}{|\nabla a_i - \nabla b_i|}\right) & \text{in } B_{r/2}(x_0), \\ \psi_i = 0 & \text{on } \partial B_{r/2}(x_0). \end{cases}$$

In other words, the functions ψ_i are obtained by the Hodge decomposition of Theorem 2.2 for the vector fields

$$-\frac{\nabla a_i - \nabla b_i}{|\nabla a_i - \nabla b_i|}$$

(extended by 0 where they are undefined). Then for any $q < \infty$, we have

$$\|\nabla\psi_i\|_{L^q(B_{r/2}(x_0))} \le C_1 r^{m/q} \tag{3.45}$$

for a constant C_1 that depends only on m and q. The Sobolev embedding theorem implies that

$$\|\psi_i\|_{L^\infty(B_{r/2}(x_0))} \le Cr. \tag{3.46}$$

Identity (3.37) and the definition of b_i imply that

$$\Delta(a_i - b_i) = \sum_{j=1}^{k} \alpha_i \cdot \omega_{ij} \quad \text{in } B_{r/2}(x_0). \tag{3.47}$$

A few integrations by parts then give

$$\begin{aligned} \int_{B_{r/2}(x_0)} |\nabla a_i - \nabla b_i|\,dx &= \int_{B_{r/2}(x_0)} (\nabla a_i - \nabla b_i) \cdot \frac{\nabla a_i - \nabla b_i}{|\nabla a_i - \nabla b_i|}\,dx \\ &= \int_{B_{r/2}(x_0)} (a_i - b_i)\Delta\psi_i\,dx \\ &= \sum_{j=1}^{k} \int_{B_{r/2}(x_0)} \alpha_j \cdot \omega_{ij}\,\psi_i\,dx \\ &= \sum_{j=1}^{k} \int_{B_{r/2}(x_0)} \langle \nabla u, e_j\rangle \cdot \omega_{ij}\,\psi_i\,dx \\ &= -\sum_{j=1}^{k} \int_{B_{r/2}(x_0)} \big\langle u - \bar{u}_{B_r(x_0)}, \nabla(\psi_i e_j)\big\rangle \cdot \omega_{ij}\,dx. \end{aligned} \tag{3.48}$$

To estimate the last term, we would like to apply Theorem 2.3 and Theorem 2.4 once more. Thus we have to extend the maps and functions in the integrand to $\mathbb{R}^m$ again. The function ψ_i can be extended by 0 outside of $B_{r/2}(x_0)$, and then it does not matter how we extend e_j. Moreover, we can replace $u - \bar{u}_{B_r(x_0)}$ by the map $\tilde{u} = \eta(u - \bar{u}_{B_r(x_0)})$, which satisfies

$$[\tilde{u}]_{\mathrm{BMO}(\mathbb{R}^m)} \leq C_2 [u]_{\mathrm{BMO}(B_r(x_0))},$$

for a constant C_2 that depends only on m and η, by Lemma 2.6. Theorem 2.3 and Theorem 2.4 now imply that

$$\begin{aligned}&\int_{B_{r/2}(x_0)} \langle u - \bar{u}_{B_r(x_0)}, \nabla(\psi_i e_j)\rangle \cdot \omega_{ij}\, dx \\ &\leq C[u]_{\mathrm{BMO}(B_r(x_0))} \left(\int_{B_r(x_0)} |\nabla(\psi_i e_j)|^2\, dx\right)^{1/2} \left(\int_{B_r(x_0)} |\omega_{ij}|^2\, dx\right)^{1/2}.\end{aligned} \tag{3.49}$$

We have

$$\int_{B_r(x_0)} |\omega_{ij}|^2\, dx \leq \int_{B_r(x_0)} |\nabla e_i|^2\, dx \leq C \int_{B_r(x_0)} |\nabla u|^2\, dx \tag{3.50}$$

by (3.34) and

$$\int_{B_r(x_0)} |\nabla(\psi_i e_j)|^2\, dx \leq C\left(r^2 \int_{B_r(x_0)} |\nabla u|^2\, dx + r^m\right) \tag{3.51}$$

by (3.45), (3.46), and (3.34). Combining (3.48) with the estimates (3.49)–(3.51), we obtain

$$\begin{aligned}&\int_{B_{r/2}(x_0)} |\nabla a_i - \nabla b_i|\, dx \\ &\leq C[u]_{\mathrm{BMO}(B_r(x_0))} \left(\int_{B_r(x_0)} |\nabla u|^2\, dx\right)^{1/2} \left(r^2 \int_{B_r(x_0)} |\nabla u|^2\, dx + r^m\right)^{1/2}.\end{aligned} \tag{3.52}$$

It remains to estimate the expression

$$\int_{B_{\theta r}(x_0)} |\nabla b_i|\, dx.$$

Being a solution of (3.44), however, the functions b_i have nice properties. We can apply the mean value formula for Laplace's equation to their gradients, and thus we find

$$\int_{B_{\theta r}(x_0)} |\nabla b_i|\,dx \le C\theta^m \int_{B_{r/2}(x_0)} |\nabla b_i|\,dx \tag{3.53}$$

for any $\theta \in (0, \frac{1}{4}]$. We now combine (3.43), (3.52), and (3.53). We obtain

$$\begin{aligned}
&\int_{B_{\theta r}(x_0)} |\nabla u|\,dx \\
&\le \sum_{i=1}^{k} \int_{B_{\theta r}(x_0)} (|A_i| + |\nabla a_i - \nabla b_i| + |\nabla b_i|)\,dx \\
&\le \sum_{i=1}^{k} \int_{B_{r/2}(x_0)} (|A_i| + |\nabla a_i - \nabla b_i| + C\theta^m |\nabla b_i|)\,dx \\
&\le C \sum_{i=1}^{k} \int_{B_{r/2}(x_0)} (|A_i| + |\nabla a_i - \nabla b_i| + \theta^m |\nabla u|)\,dx \\
&\le C[u]_{\mathrm{BMO}(B_r(x_0))} \left(\int_{B_r(x_0)} |\nabla u|^2\,dx \right)^{1/2} \left(r^2 \int_{B_r(x_0)} |\nabla u|^2\,dx + r^m \right)^{1/2} \\
&\quad + C\theta^m \int_{B_r(x_0)} |\nabla u|\,dx.
\end{aligned}$$

Dividing the inequality by $(\theta r)^{m-1}$, we obtain (3.33), and this concludes the proof. □

For the rest of this section, we follow basically the reasoning in Section 3.5. But since we have only estimates for the L^1-norm of ∇u in Lemma 3.10, not for the Dirichlet energy, we have to replace the quantity

$$r^{2-m} \int_{B_r(x_0)} |\nabla u|^2\,dx$$

by

$$M(u; x_0, r) = \sup_{B_s(x_1) \subset B_r(x_0)} \left(s^{1-m} \int_{B_s(x_1)} |\nabla u|\,dx \right).$$

The following lemma establishes a decay estimate for $M(u; x_0, r)$.

Lemma 3.11 *There exist $\epsilon_0 > 0$ and $\theta \in (0, 1)$, both depending only on m and N, such that the following holds. Suppose $u \in H^1(B_r(x_0); N)$ is a* *stationary weakly harmonic map with*

$$r^{2-m}\int_{B_r(x_0)}|\nabla u|^2\,dx \le \epsilon_0. \tag{3.54}$$

Then

$$M(u;x_0,\theta r) \le \frac{1}{2}M(u;x_0,r).$$

Proof. If (3.54) holds, the monotonicity formula (3.9) implies that

$$s^{2-m}\int_{B_s(x_1)}|\nabla u|^2\,dx \le 2^{m-2}\epsilon_0$$

for any $x_1 \in B_{r/2}(x_0)$ and $s \in (0,r/2]$. Note also that

$$[u]_{\mathrm{BMO}(B_s(x_1))} \le CM(u;x_0,r)$$

by the Poincaré inequality. Lemma 3.10, applied to the ball $B_s(x_1)$ instead of $B_r(x_0)$, yields

$$(\theta s)^{1-m}\int_{B_{\theta s}(x_1)}|\nabla u|\,dx \le C(\theta+\sqrt{\epsilon_0}+\epsilon_0)M(u;x_0,r).$$

It suffices now to choose θ and ϵ such that the factor $C(\theta+\sqrt{\epsilon_0}+\epsilon_0)$ in this inequality is at most $\frac{1}{2}$. Then we can prove the lemma by taking the supremum over all balls $B_s(x_1)$ such that $B_{\theta s}(x_1) \subset B_{\theta r}(x_0)$. □

Proposition 3.4 *There exists a number $\epsilon > 0$, depending only on m and N, such that any stationary weakly harmonic map $u \in H^1(B_r(x_0);N)$ with*

$$r^{2-m}\int_{B_r(x_0)}|\nabla u|^2\,dx \le \epsilon \tag{3.55}$$

is in $C^\infty(B_{r/2}(x_0);N)$.

Proof. The arguments are essentially the same as in the proof of Proposition 3.3. Suppose that (3.55) holds for $\epsilon = 2^{2-m}\epsilon_0$, where ϵ_0 is the constant from Lemma 3.11. Then the conditions of Lemma 3.11 are satisfied for any ball $B_s(x_1)$ with $x_1 \in B_{r/2}(x_0)$ and $s \in (0,r/2]$. Applying the lemma repeatedly, we prove that

$$M(u;x_1,s) \le C_1 s^\alpha$$

for all $x_1 \in B_{r/2}(x_0)$ and all $s \in (0, s/2]$, where $\alpha > 0$ and C_1 are numbers that do not depend on x_1 or s. In particular

$$\sup_{x_1 \in B_{r/2}(x_0)} \sup_{0<s\leq r/2} \left(s^{1-m-\alpha} \int_{B_s(x_1)} |\nabla u| \, dx \right) < \infty.$$

We now use Morrey's decay lemma (Lemma 2.1) to prove that u is Hölder continuous in $B_{r/2}(x_0)$ and we apply Theorem 3.1 to show that it is smooth in $B_{r/2}(x_0)$. □

Finally we obtain the same kind of partial regularity as in Theorem 3.2. The arguments are exactly the same here, therefore we do not repeat them.

Theorem 3.4 *Suppose $u \in H^1(\Omega; N)$ is a stationary weakly harmonic map. There exists a relatively closed set $\Sigma \in \Omega$ of vanishing $(m-2)$-dimensional Hausdorff measure, such that $u \in C^\infty(\Omega \backslash \Sigma; N)$.*

Again we observe that the condition (3.32) is unnecessary if $m = 2$. We obtain a better result in this case by the same arguments.

Theorem 3.5 *Suppose $m = 2$. Then any weakly harmonic map $u \in H^1(\Omega; N)$ is in $C^\infty(\Omega; N)$.*

3.8 Generalized Harmonic Maps

In the arguments of Section 3.5 it was important that equation (3.16) for harmonic maps into spheres can be written in the form

$$\operatorname{div}(u \wedge \nabla u) = 0 \quad \text{in } \Omega. \tag{3.56}$$

This equation is also useful for another purpose. We can use it in order to generalize the notion of harmonic maps, at least if the target manifold is the sphere $\mathbb{S}^{n-1}$. Whereas equation (3.16) requires at least $u \in H^1_{\text{loc}}(\Omega; \mathbb{S}^{n-1})$, we can make sense of (3.56) already for $u \in W^{1,1}_{\text{loc}}(\Omega; \mathbb{S}^{n-1})$. After all, the equation (in the weak sense) is equivalent to

$$\int_\Omega (u^i \nabla u^j - u^j \nabla u^i) \cdot \nabla \phi \, dx = 0, \quad i, j = 1, \ldots, n,$$

for all $\phi \in C_0^\infty(\Omega)$. This observation leads to the following definition.

Definition 3.6 A map $u \in W^{1,1}_{\text{loc}}(\Omega; \mathbb{S}^{n-1})$ is called a *generalized harmonic map* if it is a weak solution of (3.56).

Of course a generalized harmonic map that is not in $H^1(\Omega; \mathbb{S}^{n-1})$ has infinite Dirichlet energy and cannot be interpreted as a critical point of the functional E. Nevertheless, generalized harmonic maps can occur in the context of variational problems. An example is the theory of Ginzburg-Landau vortices, which was first studied by F. Bethuel, H. Brezis, and F. Hélein [1993; 1994], and which is used to model certain aspects of superconductors. In this theory, the asymptotic behavior of minimizers (or, more generally, critical points) of the functionals

$$I_\epsilon(u) = \frac{1}{2} \int_\Omega \left(|\nabla u|^2 + \frac{1}{2\epsilon^2}(1 - |u|^2)^2 \right) dx, \quad u \in H^1(\Omega; \mathbb{C}),$$

is studied, as ϵ tends to 0. We give a short description of some of the results from this theory in order to show why generalized harmonic maps play a role here.

Suppose we have a family of maps $u_\epsilon \in H^1(\Omega; \mathbb{C})$ minimizing I_ϵ for fixed boundary data (not depending on ϵ). We assume that there exists a sequence $\epsilon_k \to 0$ such that u_{ϵ_k} converge to a limit map in some sense (that is not specified any further here). The second term in the definition of I_ϵ will force the maps u_ϵ to take values close to the unit sphere $\mathbb{S}^1$ as ϵ becomes small, and any reasonable limit will have values on the 1-sphere. At the same time the minimizers try to have small Dirichlet energy, therefore we expect that the limit is a harmonic map in the space $H^1(\Omega; \mathbb{S}^1)$—provided that the boundary values admit such a map. In general the space of maps in $H^1(\Omega; \mathbb{S}^1)$ satisfying the given boundary conditions may be empty. For instance, suppose $m = 2$ and we require that $u|_{\partial\Omega} = g$ for a smooth map $g : \partial\Omega \to \mathbb{S}^1$ whose topology does not admit a continuous extension to $\overline{\Omega}$ with values in $\mathbb{S}^1$. Then there can be no harmonic map $u \in H^1(\Omega; \mathbb{S}^1)$ with these boundary data (for it would be smooth in the interior of Ω by Theorem 3.5, and similar arguments would prove that it is even smooth up to the boundary). Instead we will obtain a limit that is harmonic away from a finite set of points. At these points it exhibits vortex-like structures concentrating an infinite amount of energy. The limit map is in $W^{1,p}(\Omega; \mathbb{S}^1)$, however, for any $p < 2$, and it is a generalized harmonic map. We do not discuss this theory any further; a reader who is interested in more details is referred to F. Bethuel, H. Brezis, and F. Hélein [1994] or one of the many other works which discuss the theory.

Generalized harmonic maps may be discontinuous even for $m = 2$. A simple example of such a map is given by

$$u(x) = \frac{x}{|x|}, \quad x \in \mathbb{R}^2.$$

Obviously this map has a singularity at 0. We immediately calculate, however, that $\operatorname{div}(u \wedge \nabla u) = 0$ away from 0. Thus

$$\begin{aligned}\int_{\mathbb{R}^2} (u^1 \nabla u^2 - u^2 \nabla u^1) \cdot \nabla \phi \, dx &= \lim_{r \searrow 0} \int_{\mathbb{R}^2 \setminus B_r(0)} (u^1 \nabla u^2 - u^2 \nabla u^1) \cdot \nabla \phi \, dx \\ &= -\lim_{r \searrow 0} \int_{S_r(0)} \frac{x}{r} \cdot (u^1 \nabla u^2 - u^2 \nabla u^1) \, \phi \, dx,\end{aligned}$$

and the last limit vanishes. Hence u is a generalized harmonic map. There are also examples of generalized harmonic maps in two dimensions that are discontinuous everywhere, constructed by L. Almeida [1995].

We need thus additional conditions again if we want to prove regularity results, this time even for $m = 2$. The condition (3.8) that we used for weakly harmonic maps, however, no longer makes sense if we have maps with infinite energy. The monotonicity formula becomes meaningless likewise. The one consequence of the monotonicity formula, on the other hand, that was crucial for the regularity results, is the estimate of the BMO-seminorm it implies. We can still consider the expression $[u]_{\mathrm{BMO}(\Omega)}$ for generalized harmonic maps, thus it is natural to work with maps that are sufficiently small in the BMO-sense. We will see that such a condition can be used to show regularity for generalized harmonic maps, at least if we already know that $u \in W^{1,q}_{\mathrm{loc}}(\Omega; \mathbb{S}^{n-1})$ for some $q > 1$. The problem is open for maps that are only in $W^{1,1}_{\mathrm{loc}}(\Omega; \mathbb{S}^{n-1})$.

The arguments in this chapter are based on the same ideas as the proofs in Section 3.5, combined with additional methods. The core of these arguments was outlined in the paper R. Moser [2003a], and the arguments were improved by P. Strzelecki [2003] in order to obtain the results presented in this section. The general strategy is to show that a generalized harmonic map $u \in W^{1,p}_{\mathrm{loc}}(\Omega; \mathbb{S}^{n-1})$ with sufficiently small BMO-seminorm is even in $W^{1,\bar{p}}_{\mathrm{loc}}(\Omega, \mathbb{S}^{n-1})$ for some $\bar{p} > p$. The principal tool to achieve this is the Gehring type lemma of Section 2.6. This way one can improve the integrability stepwise until one reaches the conclusion that $u \in H^1_{\mathrm{loc}}(\Omega; \mathbb{S}^{n-1})$ after a finite number of steps. With the arguments of Section 3.5 one can then prove Hölder continuity, and finally higher regularity with the results from Section 3.4.

Proposition 3.5 *For every $q > 1$ there exist $\epsilon, \lambda > 0$ and C_0, depending only on m, n, and q, with the following property: Let $q < p \leq 2$, and suppose that $u \in W^{1,p}(B_1(0); \mathbb{S}^{n-1})$ is a generalized harmonic map satisfying*

$$[u]_{\mathrm{BMO}(B_1(0))} \leq \epsilon.$$

Then $u \in W^{1,p+\lambda}(B_{1/2}(0); \mathbb{S}^{n-1})$ and

$$\left(\int_{B_{1/2}(0)} |\nabla u|^{p+\lambda}\, dx\right)^{1/(p+\lambda)} \leq C_0 \left(\int_{B_1(0)} |\nabla u|^p\, dx\right)^{1/p}.$$

Proof. Let $B_r(x_0)$ be a ball in $B_1(0)$. We choose a cut-off function $\eta \in C_0^\infty(B_1(0))$ with $\eta \equiv 1$ in $B_{1/2}(0)$ and $|\nabla \eta| \leq 4$, and we set $\eta_{x_0,r}(x) = \eta((x - x_0)/r)$. We consider the function

$$v_{x_0,r} = \eta_{x_0,r}(u - \bar{u}_{B_r(x_0)}).$$

We use Theorem 2.1 to obtain a Hodge decomposition of the form

$$|\nabla v_{x_0,r}|^{p-2}\, u \wedge \nabla v_{x_0,r} = \nabla \phi + \Phi$$

for a function $\phi \in W^{1,p/(p-1)}(\mathbb{R}^m; \Lambda_2 \mathbb{R}^n)$ and a vector field $\Phi \in L^{p/(p-1)}(\mathbb{R}^m; \mathbb{R}^m \otimes \Lambda_2 \mathbb{R}^n)$ that satisfies $\operatorname{div} \Phi = 0$ in $\mathbb{R}^m$. We also have the estimates

$$\|\nabla \phi\|_{L^s(\mathbb{R}^m)} + \|\Phi\|_{L^s(\mathbb{R}^m)} \leq C_1 \|\nabla v_{x_0,r}\|^{p-1}_{L^{s(p-1)}(B_r(x_0))}$$

for any $s \in (1/(p-1), p/(p-1)]$, where C_1 is a constant that depends only on m, n, s, and q. Combining this inequality with the Poincaré inequality, we obtain

$$\|\nabla \phi\|^s_{L^s(\mathbb{R}^m)} + \|\Phi\|^s_{L^s(\mathbb{R}^m)} \leq C_2 \int_{B_r(x_0)} |\nabla u|^{s(p-1)}\, dx \tag{3.57}$$

for another constant C_2 that depends only on m, n, s, and q.

Now we observe that

$$|\nabla u|^p = |u \wedge \nabla u|^p = \langle u \wedge \nabla v_{x_0,r}, \nabla \phi + \Phi \rangle \quad \text{in } B_{r/2}(x_0),$$

where $\langle\,\cdot\,,\,\cdot\,\rangle$ denotes the standard scalar product in $\mathbb{R}^m \otimes \Lambda_2\mathbb{R}^n$. (We will later use the same symbol also for the scalar product in $\Lambda_2\mathbb{R}^n$.) Thus

$$
\begin{aligned}
2^{-m} \fint_{B_{r/2}(x_0)} |\nabla u|^p \, dx &= \fint_{B_r(x_0)} \langle u \wedge \nabla v_{x_0,r}, \nabla\phi\rangle \, dx \\
&\quad + \fint_{B_r(x_0)} \langle u \wedge \nabla v_{x_0,r}, \Phi\rangle \, dx \\
&= \fint_{B_r(x_0)} \nabla\eta_{x_0,r} \cdot \big\langle u \wedge (u - \bar{u}_{B_r(x_0)}), \nabla\phi\big\rangle \, dx \\
&\quad + \fint_{B_r(x_0)} \eta_{x_0,r} \, \langle u \wedge \nabla u, \nabla\phi\rangle \, dx \\
&\quad + \fint_{B_r(x_0)} \langle u \wedge \nabla v_{x_0,r}, \Phi\rangle \, dx. \qquad (3.58)
\end{aligned}
$$

We have

$$
\begin{aligned}
\int_{B_r(x_0)} \langle u \wedge \nabla v_{x_0,r}, \Phi\rangle \, dx &= \int_{B_r(x_0)} \langle v_{x_0,r} \wedge \nabla u, \Phi\rangle \, dx \\
&\leq C_3 [v_{x_0,r}]_{\mathrm{BMO}(\mathbb{R}^m)} \|\nabla u\|_{L^p(B_r(x_0))} \|\Phi\|_{L^{p/(p-1)}(\mathbb{R}^m)},
\end{aligned}
$$

for a constant C_3 that depends only on m, n, and q, by the compensated compactness result of Theorem 2.4 and the BMO-Hardy space duality of Theorem 2.3. Together with (3.57) and Lemma 2.6, this inequality implies

$$
\fint_{B_r(x_0)} \langle u \wedge \nabla v_{x_0,r}, \Phi\rangle \, dx \leq C_4 \epsilon \fint_{B_r(x_0)} |\nabla u|^p \, dx. \qquad (3.59)
$$

Here C_4 is a constant that depends only on m, n, q, and the choice of η.

We can estimate one of the terms on the right-hand side of (3.58) by the Sobolev inequality: We have

$$
\begin{aligned}
&\fint_{B_r(x_0)} \nabla\eta_{x_0,r} \cdot \big\langle u \wedge (u - \bar{u}_{B_r(x_0)}), \nabla\phi\big\rangle \, dx \\
&\quad \leq C_5 r^{-1} \left(\fint_{B_r(x_0)} |u - \bar{u}_{B_r(x_0)}|^p \, dx \right)^{1/p} \left(\fint_{B_r(x_0)} |\nabla\phi|^{\frac{p}{p-1}} \, dx \right)^{\frac{p-1}{p}} \\
&\quad \leq C_6 \left(\fint_{B_r(x_0)} |\nabla u|^{\frac{mp}{m+p}} \, dx \right)^{\frac{m+p}{mp}} \left(\fint_{B_r(x_0)} |\nabla\phi|^{\frac{p}{p-1}} \, dx \right)^{\frac{p-1}{p}} \\
&\quad \leq C_7 \left(\fint_{B_r(x_0)} |\nabla u|^{\frac{mp}{m+p}} \, dx \right)^{\frac{m+p}{mp}} \left(\fint_{B_r(x_0)} |\nabla u|^p \, dx \right)^{\frac{p-1}{p}}
\end{aligned}
$$

for constants C_5, C_6, and C_7 that depend only on m, n, and q. In the third step we have used inequality (3.57) again. It now follows from Young's inequality that for any $\delta > 0$, we have

$$\fint_{B_r(x_0)} \nabla\eta_{x_0,r} \cdot \left\langle u \wedge (u - \bar{u}_{B_r(x_0)}), \nabla\phi \right\rangle dx$$

$$\leq \delta^{p/(p-1)} \fint_{B_r(x_0)} |\nabla u|^p \, dx + \frac{C_8}{\delta^p} \left(\fint_{B_r(x_0)} |\nabla u|^{\frac{mp}{m+p}} \, dx \right)^{\frac{m+p}{m}} \tag{3.60}$$

for a constant C_8 that depends only on m, n, and q.

With an integration by parts we also obtain

$$\fint_{B_r(x_0)} \eta_{x_0,r} \langle u \wedge \nabla u, \nabla\phi \rangle \, dx = -\fint_{B_r(x_0)} \nabla\eta_{x_0,r} \left\langle u \wedge \nabla u, \phi - \bar{\phi}_{B_r(x_0)} \right\rangle dx,$$

and we can use again the Sobolev inequality and Young's inequality to show that

$$\fint_{B_r(x_0)} \eta_{x_0,r} \langle u \wedge \nabla u, \nabla\phi \rangle \, dx$$

$$\leq C_9 r^{-1} \left(\fint_{B_r(x_0)} |\nabla u|^p \, dx \right)^{1/p} \left(\fint_{B_r(x_0)} |\phi - \bar{\phi}_{B_r(x_0)}|^{\frac{p}{p-1}} \, dx \right)^{\frac{p-1}{p}}$$

$$\leq C_{10} \left(\fint_{B_r(x_0)} |\nabla u|^p \, dx \right)^{1/p} \left(\fint_{B_r(x_0)} |\nabla\phi|^{\frac{mp}{m(p-1)+p}} \, dx \right)^{\frac{m(p-1)+p}{mp}}$$

$$\leq \delta^p \fint_{B_r(x_0)} |\nabla u|^p \, dx + \frac{C_{11}}{\delta^{p/(p-1)}} \left(\fint_{B_r(x_0)} |\nabla\phi|^{\frac{mp}{m(p-1)+p}} \, dx \right)^{\frac{m(p-1)+p}{m(p-1)}}$$

for any $\delta > 0$ and for constants C_9, C_{10}, and C_{11} that depend only on m, n, and q. We apply inequality (3.57) once more, this time for

$$s = \frac{mp}{m(p-1)+p}.$$

We finally obtain

$$\fint_{B_r(x_0)} \eta_{x_0,r} \langle u \wedge \nabla u, \nabla\phi \rangle \, dx$$

$$\leq \delta^p \fint_{B_r(x_0)} |\nabla u|^p \, dx + \frac{C_{12}}{\delta^{p/(p-1)}} \left(\fint_{B_r(x_0)} |\nabla u|^{\frac{mp}{m+p/(p-1)}} \, dx \right)^{\frac{m(p-1)+p}{m(p-1)}} \tag{3.61}$$

for any $\delta > 0$ and for a constant C_{12} which depends only on m, n, and q.

Combining (3.58)–(3.61), and using the Hölder inequality to estimate the right-hand side of (3.61) further, we find that

$$\fint_{B_{r/2}(x_0)} |\nabla u|^p \, dx \leq (2\delta^p + C_4\epsilon) \fint_{B_r(x_0)} |\nabla u|^p \, dx$$
$$+ C_{13}\delta^{p/(1-p)} \left(\fint_{B_r(x_0)} |\nabla u|^{\frac{mp}{m+p}} \, dx \right)^{\frac{m+p}{m}},$$

where $\delta > 0$ can be chosen arbitrarily and C_{13} is a constant that depends only on m, n, and q. This is true for every ball $B_r(x_0) \subset B_1(0)$. If ϵ and δ are chosen sufficiently small, then we can apply Lemma 2.11 to the function

$$|\nabla u|^{\frac{mp}{m+p}}.$$

The gain of integrability that the lemma implies depends on q, but not on p. We conclude that the claim of the proposition is true. □

To prove regularity of generalized harmonic maps which are small in the BMO-sense, we can now simply apply Proposition 3.5 repeatedly. Suppose $u \in W^{1,q}_{\mathrm{loc}}(\Omega; \mathbb{S}^{n-1})$ is a solution of (3.56), where $q > 1$. Suppose further that

$$[u]_{\mathrm{BMO}(\Omega)} \leq \epsilon$$

for the number ϵ from Proposition 3.5 (belonging to the given q).

For any given $x_0 \in \Omega$, we can choose a radius $r > 0$ such that $B_r(x_0) \subset \Omega$ and consider the rescaled map

$$v(x) = u(rx + x_0).$$

This is still a generalized harmonic map and satisfies the conditions of the proposition, hence

$$v \in W^{1,q+\lambda}(B_{1/2}(0); \mathbb{S}^{n-1}).$$

We conclude

$$u \in W^{1,q+\lambda}_{\mathrm{loc}}(\Omega; \mathbb{S}^{n-1}).$$

We apply the same arguments again and obtain

$$u \in W^{1,q+2\lambda}_{\mathrm{loc}}(\Omega; \mathbb{S}^{n-1}).$$

Repeating the procedure, we reach in finitely many steps the conclusion

$$u \in H^1_{\mathrm{loc}}(\Omega; \mathbb{S}^{n-1}).$$

Hence u is actually a weakly harmonic map, not merely a generalized harmonic map. Next we observe that we can apply the arguments from Section 3.5 if we know that $[u]_{\mathrm{BMO}(\Omega)}$ is sufficiently small, and these arguments imply Hölder continuity of u. Higher regularity then follows from Theorem 3.1. Thus we have proved the following result.

Theorem 3.6 *For any $q > 1$ there exists a number $\epsilon > 0$ such that any generalized harmonic map $u \in W^{1,q}_{\mathrm{loc}}(\Omega; \mathbb{S}^{n-1})$ satisfying*

$$[u]_{\mathrm{BMO}(\Omega)} \leq \epsilon,$$

is in $C^\infty(\Omega; \mathbb{S}^{n-1})$.

Remark 3.4 *There is an alternative to using the results of Sections 3.5 and 3.4 in the arguments above. Indeed we can continue to improve the integrability of u with the arguments of this section until we reach the conclusion that*

$$u \in W^{1,m+\alpha}_{\mathrm{loc}}(\Omega; \mathbb{S}^{n-1})$$

for a number $\alpha > 0$. Then the Sobolev embedding theorem implies that

$$u \in C^{1,\alpha/(m+\alpha)}_{\mathrm{loc}}(\Omega; \mathbb{S}^{n-1}).$$

This is enough to derive higher regularity with easy bootstrap-arguments (like in the proof of Theorem 3.1). With this method we also have an alternative proof for the regularity of weakly harmonic maps into spheres that are small in the BMO*-sense. This is another observation that has been made by P. Strzelecki [2003].*

3.9 Further Remarks

3.9.1 *Homogeneous Manifolds*

The arguments for the target manifold $\mathbb{S}^{n-1}$ in Section 3.5 are easier than for general targets, because the symmetry of the sphere permits to rewrite the harmonic map equation (3.16) in the form

$$\operatorname{div}(u \wedge \nabla u) = 0 \quad \text{in } \Omega. \tag{3.62}$$

For the same reason, we were able to generalize the notion of harmonic maps in Section 3.8. The sphere is not the only manifold, however, that has this property. In this section we give a brief description of the results which lead to a similar reformulation of the harmonic map equation for certain other targets. We do not go into details here. For further information, the reader is referred to F. Hélein [1991b].

Suppose there exists a Lie group G of dimension K that acts transitively on N by isometries. (If such a Lie group exists, N is called a homogeneous manifold.) It was shown by F. Hélein [1991b] that under these conditions, there exist $2K$ smooth tangent vector fields $X_1, \ldots, X_K, Y_1, \ldots, Y_K$ on N, such that

- each X_l is a Killing vector field, i. e.,

$$\left\langle D_Y^N X_l(y), Z\right\rangle + \left\langle Y, D_Z^N X_l(y)\right\rangle = 0 \tag{3.63}$$

  for every $y \in N$ and all $Y, Z \in T_y N$, and
- every tangent vector $Z \in T_y N$ satisfies

$$Z = \sum_{l=1}^{k} \langle Z, X_l(y)\rangle\, Y_l(y).$$

For instance, there exists such a Lie group for the sphere $\mathbb{S}^{n-1}$, namely $\mathrm{SO}(n)$. Its dimension is $n(n-1)/2$. For this manifold, the vector fields

$$X_{ij}(y) = y^i e_j - y^j e_i, \quad 1 \le i < j \le n,$$

and

$$Y_{ij}(y) = y^i e_j - y^i y^j y, \quad 1 \le i < j \le n,$$

have the desired properties, where e_i denotes the i-th standard unit vector in $\mathbb{R}^n$.

If $u \in H^1(\Omega; N)$, it follows that

$$\nabla u = \sum_{l=1}^{K} \langle \nabla u, X_l(u)\rangle\, Y_l(u).$$

Moreover, the condition (3.63) implies

$$\operatorname{div} \langle \nabla u, X_l(u)\rangle = \langle \Delta u, X_l(u)\rangle\,, \quad l = 1, \ldots, K,$$

provided that $\Delta u \in L^1(\Omega; \mathbb{R}^n)$. If u is weakly harmonic, we have

$$\operatorname{div} \langle \nabla u, X_l(u) \rangle = 0 \quad \text{in } \Omega \tag{3.64}$$

for $l = 1, \ldots, K$.

Equation (3.64) is the equivalent of (3.62) in the more general situation. Because the vector fields X_l and Y_l are smooth, there exists a constant C, depending only on N, such that

$$\frac{1}{C}|\nabla u|^2 \leq \sum_{l=1}^{K} |\langle \nabla u, X_l(u) \rangle|^2 \leq C|\nabla u|^2.$$

Quantities of the form

$$\int_{B_r(x_0)} |\langle \nabla u, X_l(u) \rangle|^2 \, dx$$

can be estimated with the arguments in the proof of Lemma 3.8. With this method, we obtain an alternative proof for the partial regularity of stationary weakly harmonic maps into homogeneous target manifolds. This can be useful not only because the arguments of Section 3.5 are easier than the ones of Section 3.7, but also because they permit certain generalizations that are not possible for general target manifolds. We will see an example of this kind in Chapter 5.

Moreover, (3.64) can be used in order to define generalized harmonic maps. The arguments of Section 3.8 still work in the situation we consider here, thus we obtain the same regularity results. We leave it to the reader to work out the details.

3.9.2 *Energy Minimizing Maps*

Another remark is concerned with the regularity of energy minimizing maps. If $u \in H^1(\Omega; N)$ has the property that $E(u) \leq E(v)$ for every $v \in H^1(\Omega; N)$ that agrees with u outside a compact subset of Ω, then we have obviously

$$\left.\frac{d}{ds}\right|_{s=0} E(\pi_N \circ (u + s\phi)) = 0$$

for any $\phi \in C_0^\infty(\Omega; \mathbb{R}^n)$, and also

$$\left.\frac{d}{ds}\right|_{s=0} E(u(x + s\psi(x))) = 0$$

for any $\psi \in C_0^\infty(\Omega; \mathbb{R}^m)$. That is, u is a stationary weakly harmonic map. Such energy minimizing maps are obtained, for example, when a solution of the harmonic map equation (3.5) is constructed with the direct method in the calculus of variations. The condition that u is energy minimizing, however, is stronger than (3.5) and (3.8), and this fact has certain consequences for the regularity of u.

Firstly, there are easier arguments than the ones presented in this chapter to prove partial regularity results for energy minimizers. Instead of using the harmonic map equation, one can construct appropriate comparison maps and use directly the fact that u minimizes the energy. Moreover, a careful analysis of the limit of rescaled maps reveals that the singular set of energy minimizers is at most $(m-3)$-dimensional.

The arguments that can be used to prove these results are quite different from the methods we use in this chapter, and we do not explain them further. The reader can find them, e. g., in a book by L. Simon [1996].

Chapter 4

Almost Harmonic Maps

In this chapter we consider perturbations of the harmonic map equation of the form

$$\Delta u + A(u)(\nabla u, \nabla u) = f \quad \text{in } \Omega, \tag{4.1}$$

where $f : \Omega \to \mathbb{R}^n$ is a function in $L^p(\Omega; N)$ for an appropriate exponent p. We call a solution of (4.1) an almost harmonic map. We study the regularity of such solutions. Of course the regularity results we can expect depend on the exponent p. We consider three cases: First, we study the case $p > \frac{m}{2}$ (and for convenience we also assume that $p \geq 2$). In this situation, we derive Hölder continuity of solutions of (4.1) under suitable conditions. In the borderline case $p = \frac{m}{2}$, we cannot expect continuity of the solutions, but there is still an energy inequality that will prove useful. Finally, we examine the regularity of solutions of (4.1) for $p = 2$ and $m \leq 4$. In small dimensions, the problem is easier, and we can prove more than in higher dimensions. We will obtain in particular estimates for the second derivatives of solutions. Most of the results we use in this chapter are from the papers R. Moser [2003b; 2004]. The arguments are generalizations of the ones from the previous chapter, combined with a few additional arguments.

4.1 The Monotonicity Formula

We need a generalization of the monotonicity formula for stationary weakly harmonic maps if we want to prove regularity results for almost harmonic maps. To find the right generalization of the condition (3.8), we first observe that any smooth map $u \in C^\infty(\Omega; N)$ satisfies

$$\operatorname{div}\left(|\nabla u|^2 e_\alpha - 2\left\langle \frac{\partial u}{\partial x^\alpha}, \nabla u \right\rangle\right) = -2\left\langle \frac{\partial u}{\partial x^\alpha}, \Delta u \right\rangle = -2\left\langle \frac{\partial u}{\partial x^\alpha}, \tau(u) \right\rangle$$

for $\alpha = 1, \ldots, m$, where

$$\tau(u) = \Delta u + A(u)(\nabla u, \nabla u)$$

is the tension field. Thus it is natural to replace (3.8) by the following condition.

Definition 4.1 A solution $u \in H^1(\Omega; N)$ of (4.1) is called *stationary* if it satisfies

$$\operatorname{div}\left(|\nabla u|^2 e_\alpha - 2\left\langle \frac{\partial u}{\partial x^\alpha}, \nabla u \right\rangle\right) + 2\left\langle \frac{\partial u}{\partial x^\alpha}, f \right\rangle = 0 \quad \text{in } \Omega \tag{4.2}$$

for $\alpha = 1, \ldots, m$ in the distribution sense.

For stationary almost harmonic maps, we have the following version of the monotonicity formula.

Lemma 4.1 *Suppose $m \geq 3$. Let $u \in H^1(\Omega; N)$ be a map satisfying (4.2), where $f \in L^2_{\mathrm{loc}}(\Omega; \mathbb{R}^n)$. For $x_0 \in \Omega$ and $0 < \rho \leq \operatorname{dist}(x_0, \partial\Omega)$, define*

$$\Phi_f(u; x_0, \rho) = \rho^{2-m} \int_{B_\rho(x_0)} \left(|\nabla u|^2 - \frac{2}{m-2}\langle (x - x_0) \cdot \nabla u, f \rangle\right) dx.$$

Suppose $0 < s \leq r \leq \operatorname{dist}(x_0, \partial\Omega)$. Then

$$\begin{aligned} &\Phi_f(u; x_0, r) - \Phi_f(u; x_0, s) \\ &= 2\int_{B_r(x_0)\setminus B_s(x_0)} \left(\frac{|(x - x_0)\cdot \nabla u|^2}{|x - x_0|^m} - \frac{\langle (x - x_0)\cdot \nabla u, f\rangle}{(m-2)|x - x_0|^{m-2}}\right) dx. \end{aligned} \tag{4.3}$$

Proof. The arguments are very similar to the ones for the proof of Lemma 3.3. Suppose $x_0 = 0$ for simplicity. Testing (4.2) with $\chi_{B_\rho(0)} x^\alpha$, where $\chi_{B_\rho(0)}$ is the characteristic function of $B_\rho(0)$, and summing over α, we obtain

$$\begin{aligned} (m-2)\int_{B_\rho(0)} &\left(|\nabla u|^2 - \frac{2}{m-2}\langle x \cdot \nabla u, f\rangle\right) dx \\ &= \int_{S_\rho(0)} \left(\rho|\nabla u|^2 - \frac{2}{\rho}|x \cdot \nabla u|^2\right) do. \end{aligned}$$

Thus

$$\frac{d}{d\rho}\Phi_f(u;0,\rho) = 2\rho^{-m}\int_{S_\rho(0)} |x\cdot\nabla u|^2\,do - \frac{2\rho^{2-m}}{m-2}\int_{S_\rho(0)} \langle x\cdot\nabla u, f\rangle\,do.$$

Integrating this identity over the interval $[s,r)$ yields (4.3). □

The following is a consequence of Lemma 4.1. We usually refer to this as the monotonicity inequality.

Lemma 4.2 *Suppose $m \geq 4$ and $p > \frac{m}{2}$, or $m \leq 4$ and $p = 2$. Then there exists a constant C_0, depending only on m and p, such that the following holds. Let $u \in H^1(\Omega; N)$ be a solution of (4.2) for some $f \in L^p_{\mathrm{loc}}(\Omega;\mathbb{R}^n)$. If $B_s(x_0) \subset B_r(x_0) \subset \Omega$ and $\delta > 0$, then*

$$s^{2-m}\int_{B_s(x_0)} |\nabla u|^2\,dx \leq (1+\delta)r^{2-m}\int_{B_r(x_0)} |\nabla u|^2\,dx + C_0\left(1+\frac{1}{\delta}\right)\left(r^{2p-m}\int_{B_r(x_0)} |f|^p\,dx\right)^{2/p}. \quad (4.4)$$

Proof. For $m = 2$ there is nothing to prove. Thus we assume $m \geq 3$. We combine identity (4.3) with the estimates

$$\begin{aligned} &2\rho^{2-m}\left|\int_{B_\rho(x_0)} \langle (x-x_0)\cdot\nabla u, f\rangle\,dx\right| \\ &\quad\leq \delta'\rho^{2-m}\int_{B_r(x_0)} |\nabla u|^2\,dx + \frac{\rho^{4-m}}{\delta'}\int_{B_\rho(x_0)} |f|^2\,dx \\ &\quad\leq \delta'\rho^{2-m}\int_{B_r(x_0)} |\nabla u|^2\,dx + \frac{\omega_m^{(p-2)/p}}{\delta'}\left(r^{2p-m}\int_{B_r(x_0)} |f|^p\,dx\right)^{2/p} \end{aligned}$$

for any $\delta' > 0$ and any $\rho \in (0,r]$, and

$$\begin{aligned} &2\left|\int_{B_r(x_0)\setminus B_s(x_0)} \frac{\langle (x-x_0)\cdot\nabla u, f\rangle}{|x-x_0|^{m-2}}\,dx\right| \\ &\quad\leq \int_{B_r(x_0)\setminus B_s(x_0)} \frac{|(x-x_0)\cdot\nabla u|^2}{|x-x_0|^m}\,dx + \int_{B_r(x_0)\setminus B_s(x_0)} \frac{|f|^2}{|x-x_0|^{m-4}}\,dx. \end{aligned}$$

If $m \le 4$ and $p = 2$, inequality (4.4) follows immediately. If $m \ge 4$ and $p > \frac{m}{2}$, we have

$$\begin{aligned}
&\int_{B_r(x_0)\setminus B_s(x_0)} \frac{|f|^2}{|x-x_0|^{m-4}}\,dx \\
&\quad = \int_s^r \rho^{4-m} \int_{S_\rho(x_0)} |f|^2\,do\,d\rho \\
&\quad = r^{4-m} \int_{B_r(x_0)} |f|^2\,dx - s^{4-m} \int_{B_s(x_0)} |f|^2\,dx \\
&\qquad + (m-4) \int_s^r \rho^{3-m} \int_{B_\rho(x_0)} |f|^2\,dx\,d\rho \\
&\quad \le \omega_m^{(p-2)/p} \left(r^{2p-m} \int_{B_r(x_0)} |f|^p\,dx \right)^{2/p} \\
&\qquad + (m-4)\omega_m^{(p-2)/p} \int_s^r \rho^{3-2m/p} \left(\int_{B_\rho(x_0)} |f|^p\,dx \right)^{2/p} d\rho \\
&\quad \le \left(1 + \frac{m-4}{4-2m/p}\right) \omega_m^{(p-2)/p} \left(r^{2p-m} \int_{B_r(x_0)} |f|^p\,dx \right)^{2/p}.
\end{aligned}$$

We combine the estimates, and the proof is finished. □

4.2 Hölder Continuity

Let now $u \in H^1(\Omega; N)$ be a stationary solution of (4.1) for a function $f \in L^p(\Omega; \mathbb{R}^n)$ with $p > \frac{m}{2}$ and $p \ge 2$. We want to show that u is Hölder continuous. The arguments we use in this situation are a direct generalization of the arguments of Section 3.7. The key lemma of that section, Lemma 3.10, is replaced by the following inequality.

Lemma 4.3 *There exists a constant C, depending only on m and N, such that any solution $u \in H^1(B_r(x_0); N)$ of (4.1) with $f \in L^{m/2}(B_r(x_0); \mathbb{R}^n)$ and*

$$r^{2-m} \int_{B_r(x_0)} |\nabla u|^2\,dx \le \epsilon,$$

satisfies

$$(\theta r)^{1-m}\int_{B_{\theta r}(x_0)}|\nabla u|\,dx \le C(\sqrt{\epsilon}+\epsilon)[u]_{\mathrm{BMO}(B_r(x_0))}+C\|f\|_{L^{m/2}(B_r(x_0))}$$
$$+\,C\theta r^{1-m}\int_{B_r(x_0)}|\nabla u|\,dx \tag{4.5}$$

for any $\theta \in (0,1)$.

Proof. By the same reasoning as in Section 3.6, we can assume that there exists an orthonormal tangent frame $(e_1,\ldots,e_k)$ along u that satisfies (3.34), such that the vector fields ω_{ij}, defined by (3.29), have the property (3.30). Then we can use exactly the same arguments as in the proof of Lemma 3.10, except that we have to add certain terms in some of the formulas. Instead of (3.37), we now have

$$\Delta a_i = \sum_{j=1}^{k}\alpha_j\cdot\omega_{ij}+\langle f,e_i\rangle \quad \text{in } B_{r/2}(x_0).$$

Identity (3.47) then becomes

$$\Delta(a_i-b_i) = \sum_{j=1}^{k}\alpha_i\cdot\omega_{ij}+\langle f,e_i\rangle \quad \text{in } B_{r/2}(x_0),$$

and the term

$$\int_{B_{r/2}(x_0)}\langle f,e_i\rangle\,\psi_i\,dx$$

has to be added in the last two lines of (3.48). We can estimate

$$\int_{B_{r/2}(x_0)}\langle f,e_i\rangle\,\psi_i\,dx \le C\|f\|_{L^{m/2}(B_r(x_0))}r^{m-1}, \tag{4.6}$$

using estimate (3.46) for ψ_i. Finally, the right-hand side of (4.6) contributes to the estimates in the way indicated in (4.5). □

Proposition 4.1 *Suppose* $p\ge 2$ *and* $p>\frac{m}{2}$. *There exist two numbers* $\epsilon,\alpha>0$, *depending only on* m, N, *and* p, *such that the following holds. Suppose* $u\in H^1(B_r(x_0);N)$ *is a stationary solution of (4.1) with* $f\in L^p(B_r(x_0);\mathbb{R}^n)$ *and*

$$r^{2-m}\int_{B_r(x_0)}|\nabla u|^2\,dx\le\epsilon.$$

Then $u \in C^{0,\alpha}(B_{r/2}(x_0); N)$.

Proof. For simplicity we work with $B_1(0)$ instead of $B_r(x_0)$. By Lemma 4.2, we have

$$r^{2-m} \int_{B_r(x_0)} |\nabla u|^2 \, dx \le C\epsilon + C_1 r^{4-2m/p} \|f\|^2_{L^p(B_r(x_0))}$$

for any $x_0 \in B_{3/4}(0)$ and any $r \in (0, \frac{1}{4}]$, where C_1 is a constant that depends only on m, N, and p. Thus for any fixed $\epsilon_0 > 0$, there exists a number $r_0 > 0$ such that

$$r^{2-m} \int_{B_r(x_0)} |\nabla u|^2 \, dx \le \epsilon_0$$

for any $x_0 \in B_{3/4}(0)$ and any $r \in (0, r_0]$, provided that ϵ is sufficiently small. Define

$$M(u; x_0, r) = \sup_{B_s(x_1) \subset B_r(x_0)} \left(s^{1-m} \int_{B_s(x_1)} |\nabla u| \, dx \right),$$

so that we have

$$[u]_{\mathrm{BMO}(B_r(x_0))} \le C M(u; x_0, r)$$

by the Poincaré inequality. Applying Lemma 4.3 for a sufficiently small number $\theta > 0$ and noting that

$$\|f\|_{L^{m/2}(B_r(x_0))} \le C_2 r^{(2p-m)/p} \|f\|_{L^p(B_r(x_0))}$$

for a constant C_2 that depends only on m, N, and p, we can draw the following conclusion: If ϵ_0 is small enough, then

$$M(u; x_0, \theta r) \le \frac{1}{2} M(u; x_0, r) + C_3 r^{(2p-m)/p} \|f\|_{L^p(B_1(0))}$$

for any $x_0 \in B_{1/2}(0)$ and any $r \in (0, r_0]$, where C_3 is a constant that depends only on m, N, and p. It follows that there exists a number $\alpha > 0$ such that

$$\sup_{x_0 \in B_{1/2}(0)} \sup_{0 < r \le r_0} \left(r^{1-m-\alpha} \int_{B_r(x_0)} |\nabla u| \, dx \right) < \infty.$$

Lemma 2.1 then implies $u \in C^{0,\alpha}(B_{1/2}(0))$. □

Like in the previous chapters, we can combine the proposition with a covering argument to show that a stationary solution of (4.1) for an exponent p satisfying $p \geq 2$ and $p > \frac{m}{2}$, is Hölder continuous away from a small set.

Theorem 4.1 *Suppose $u \in H^1(\Omega; N)$ is a stationary solution of (4.1) for a function $f \in L^p(\Omega; \mathbb{R}^n)$, where $p > \frac{m}{2}$ and $p \geq 2$. Then there exists a relatively closed set $\Sigma \subset \Omega$ of vanishing $(m-2)$-dimensional Hausdorff measure, such that $u \in C^{0,\alpha}_{\text{loc}}(\Omega \backslash \Sigma; N)$ for a number $\alpha > 0$ that depends only on m, N, and p.*

Remark 4.1 *In the proof of Theorem 4.1 we have only used the fact that f satisfies an estimate of the form*

$$\int_{B_r(x_0)} |f|^2 \, dx \leq C r^{m-4+\alpha}$$

for all balls $B_r(x_0) \subset \Omega$ and for certain numbers $\alpha > 0$ and C. Thus the result of the theorem still holds if the condition $f \in L^p(\Omega; \mathbb{R}^m)$ is replaced with this inequality. We leave it to the reader to verify the details.

For almost harmonic maps, continuity does not imply smoothness in general. But if u is the solution of a special variational problem, higher regularity may follow from continuity. This is true, e. g., for critical points of many functionals obtained from the Dirichlet energy by adding terms of lower order. For instance, it is not difficult to prove that continuous critical points of the functional F discussed in the introduction are smooth (indeed, the same arguments as in Section 3.4 can be used here). We do not discuss higher regularity in detail, however.

4.3 A Lemma of S. Luckhaus

Before we continue to examine almost harmonic maps, we need a few additional tools. In this section we present a lemma that is concerned with the extension of a map that is defined on the boundary of an annulus $B_1(0) \backslash B_{1-\lambda}(0)$, into the interior of this annulus. It is a version of a result due to S. Luckhaus [1988].

Lemma 4.4 *There exists a constant C, depending only on m and n, such that the following holds. Suppose $K \subset \mathbb{R}^n$ is a closed set, $u, v \in H^1(S_1(0), K)$, and $\lambda \in (0, \frac{1}{2}]$. Then there exists some $\phi \in$*

$H^1(B_1(0)\backslash\overline{B_{1-\lambda}(0)};\mathbb{R}^n)$ that coincides with u on $S_1(0)$ and with $v(x/(1-\lambda))$ on $S_{1-\lambda}(0)$ in the sense of traces, such that

$$\int_{B_1(0)\backslash B_{1-\lambda}(0)} |\nabla\phi|^2\,dx \le C\lambda\int_{S_1(0)}(|\nabla u|^2+|\nabla v|^2)\,do + \frac{C}{\lambda}\int_{S_1(0)}|u-v|^2\,do, \tag{4.7}$$

and

$$\begin{aligned}\operatorname{dist}^2(\phi(x),K) &\le C\lambda^{2-m}\left(\int_{S_1(0)}(|\nabla u|^2+|\nabla v|^2)\,do\right)^{1/2}\left(\int_{S_1(0)}|u-v|^2\,do\right)^{1/2}\\ &\quad + C\lambda^{1-m}\int_{S_1(0)}|u-v|^2\,do\end{aligned} \tag{4.8}$$

for almost every $x\in B_1(0)\backslash\overline{B_{1-\lambda}(0)}$.

Remark 4.2 *Suppose $B_r(x_0)\subset\mathbb{R}^m$ is a ball with $0\in\overline{B_r(x_0)}$ and $r\ge 1$. If $u,v\in H^1(S_1(0)\cap B_r(x_0);K)$ and $\lambda\in(0,\frac{1}{2}]$ are given, then a map $\phi\in H^1(B_1(0)\backslash\overline{B_{1-\lambda}(0)};\mathbb{R}^n)$ can be found such that*

$$\phi|_{S_1(0)\cap B_r(x_0)} = u|_{S_1(0)\cap B_r(x_0)}$$

and

$$\phi|_{S_{1-\lambda}(0)\cap B_r(x_0)} = v(x/(1-\lambda))|_{S_{1-\lambda}(0)\cap B_r(x_0)}$$

in the sense of traces, and such that the inequalities (4.7) and (4.8) hold for $B_1(0)\cap B_r(x_0)\backslash B_{1-\lambda}(0)$ instead of $B_1(0)\backslash B_{1-\lambda}(0)$ and $S_1(0)\cap B_r(x_0)$ instead of $S_1(0)$. To prove this claim, we simply extend u and v appropriately and apply Lemma 4.4.

Proof. We first introduce a cell decomposition of $S_1(0)$ consisting of cells that are Lipschitz equivalent to balls of radius λ; that is, there exists a constant C, depending only on m, such that every j-cell q of the decomposition is mapped by a bi-Lipschitz map Φ_q onto the ball $\overline{B^j_\lambda(0)}\subset\mathbb{R}^j$, and the Lipschitz constants of both Φ_q and Φ_q^{-1} are bounded by C. One way to obtain such a decomposition is the following: Let $k_0\in\mathbb{Z}$ be the unique number such that

$$2^{k_0}\le\lambda<2^{k_0+1}.$$

Decompose the cube $[-1,1]^m \subset \mathbb{R}^m$ into $2^{(1-k_0)m}$ closed cubes of side length 2^{k_0} by a repeated bisection of the sides. (That is, the vertices of these cubes are points of the lattice $2^{k_0}\mathbb{Z}^m$.) Let $\tilde{Q}_j$ be the collection of all (closed) j-faces of these cubes contained in the boundary $\partial[-1,1]^m$, $j = 0, \dots, m-1$. Then $\tilde{Q}_0, \dots, \tilde{Q}_{m-1}$ give a cell decomposition of $\partial[-1,1]^m$, consisting of cells with diameters of magnitude λ. We then obtain a similar decomposition of $S_1(0)$ via the bi-Lipschitz mapping

$$\Phi(x) = \frac{x}{|x|}, \quad x \in \partial[-1,1]^m.$$

That is, we define

$$\hat{Q}_j = \left\{\Phi(q) : \ q \in \tilde{Q}_j\right\}, \quad j = 0, \dots, m-1.$$

Then the cell decomposition given by $\hat{Q}_0, \dots, \hat{Q}_{m-1}$ has the desired properties. For any rotation $R \in \mathrm{SO}(m)$, the collections

$$R\hat{Q}_j = \left\{Rq : \ q \in \hat{Q}_j\right\}, \quad j = 0, \dots, m-1,$$

give another such cell decomposition.

For convenience we use the abbreviation $|\cdot|$ for the $(m(m-1)/2)$-dimensional Hausdorff measure on $\mathrm{SO}(m)$ in the rest of the proof. We claim that there exists a constant C, depending only on m, such that for any $j = 0, \dots, m-1$, there is a set $\Lambda_j \subset \mathrm{SO}(m)$ of measure

$$|\Lambda_j| \le \frac{|\mathrm{SO}(m)|}{2m}$$

with the following property: For any $R \in \mathrm{SO}(m)\backslash\Lambda_j$, the inequalities

$$\sum_{q \in R\hat{Q}_j} \int_q (|\nabla u|^2 + |\nabla v|^2)\, dh^j \le C\lambda^{j-m+1} \int_{S_1(0)} (|\nabla u|^2 + |\nabla v|^2)\, do \qquad (4.9)$$

and

$$\sum_{q \in R\hat{Q}_j} \int_q |u - v|^2\, dh^j \le C\lambda^{j-m+1} \int_{S_1(0)} |u - v|^2\, do \qquad (4.10)$$

hold. (Remember that h^j denotes the j-dimensional Hausdorff measure.)

To prove the claim, we fist observe that for any pair of points $x, y \in S_1(0)$, there exists some $R_0 \in \mathrm{SO}(m)$ such that $R_0 x = y$. Since the mapping

$R \mapsto RR_0$ preserves the measure on $\mathrm{SO}(m)$, we have

$$\int_{\mathrm{SO}(m)} F(Rx)\,dR = \int_{\mathrm{SO}(m)} F(Ry)\,dR$$

for any integrable function F on $S_1(0)$. Here dR indicates integration with respect to the $(m(m-1)/2)$-dimensional Hausdorff measure. For similar reasons,

$$\int_{S_1(0)} F\,do = \int_{S_1(0)} F \circ R\,do$$

for any $R \in \mathrm{SO}(m)$. Thus by Fubini's theorem,

$$\begin{aligned}\int_{S_1(0)} F\,do &= \frac{1}{|\mathrm{SO}(m)|}\int_{\mathrm{SO}(m)}\int_{S_1(0)} F \circ R\,do\,dR \\ &= \frac{h^{m-1}(S_1(0))}{|\mathrm{SO}(m)|}\int_{\mathrm{SO}(m)} F(Rx_0)\,dR\end{aligned}$$

for any $x_0 \in S_1(0)$. In particular

$$\begin{aligned}\int_{\mathrm{SO}(m)}\int_{Rq} F\,dh^j\,dR &= \int_q\int_{\mathrm{SO}(m)} F(Rx)\,dR\,dh^j(x) \\ &= \frac{h^j(q)|\mathrm{SO}(m)|}{h^{m-1}(S_1(0))}\int_{S_1(0)} F\,do \qquad (4.11)\end{aligned}$$

for any $q \in \hat{Q}_j$ and any $j = 0, \ldots, m-1$. Suppose c_0 is a given constant. If the set

$$\Lambda_{F,j} = \left\{ R \in \mathrm{SO}(m) : \sum_{q \in R\hat{Q}_j}\int_q F\,dh^j > c_0\lambda^{j-m+1}\int_{S_1(0)} F\,do \right\}$$

had a measure

$$|\Lambda_{F,j}| > \frac{|\mathrm{SO}(m)|}{4m},$$

then we would find

$$\begin{aligned}\sum_{q\in\hat{Q}_j}\int_{SO(m)}\int_{Rq} F\,dh^j\,dR &\geq \sum_{q\in\hat{Q}_j}\int_{\Lambda_{F,j}}\int_{Rq} F\,dh^j\,dR \\ &= \int_{\Lambda_{F,j}}\sum_{q\in R\hat{Q}_j}\int_q F\,dh^j\,dR \\ &> \frac{c_0|SO(m)|}{4m}\lambda^{j-m+1}\int_{S_1(0)} F\,do.\end{aligned}$$

If c_0 is chosen sufficiently large, this would be in contradiction to (4.11), thus we conclude

$$|\Lambda_{F,j}| > \frac{|SO(m)|}{4m}.$$

Using these arguments twice, first for $F = |\nabla u|^2 + |\nabla v|^2$ and then for $F = |u-v|^2$, we can prove the existence of the sets Λ_j with the claimed properties.

The union of the sets $\Lambda_0, \dots, \Lambda_{m-1}$ has a measure no bigger than $|SO(m)|/2$. Thus there exists some $R \in SO(m)$ such that (4.9) and (4.10) hold for every $j = 0, \dots, m-1$. We fix some R with these properties, and then we define

$$Q_j = R\hat{Q}_j = \left\{Rq:\ q\in\hat{Q}_j\right\}, \quad j = 0,\dots,m-1.$$

Note that every $q \in Q_j$ is a j-dimensional Lipschitz manifold. Thus we can consider the gradient of functions on q with respect to the metric inherited from $\mathbb{R}^m$. We denote this gradient operator by ∇^q. We can assume that $\nabla^q u$ and $\nabla^q v$ are the projections of ∇u and ∇v, respectively, onto the appropriate tangent spaces h^j-almost everywhere on q (because this is true for almost every R).

For every $q \in Q_j$, we also consider the set

$$\bar{q} = \bigcup_{1-\lambda\leq s\leq 1} sq = \{sx:\ 1-\lambda\leq s\leq 1,\ x\in q\}.$$

We set

$$P_{j+1} = \{\bar{q}:\ q\in Q_j\}, \quad j=0,\dots,m-1.$$

The collections $P_1, \dots, P_{m-1}$ give a cell decomposition of the annulus $\overline{B_1(0)}\backslash B_{1-\lambda}(0)$. For every $p \in P_j$ there exists a bi-Lipschitz mapping Φ_p

that maps p onto $\overline{B^j_\lambda(0)}$, and the Lipschitz constants of Φ_p and Φ_p^{-1} are bounded by a constant C that depends only on m.

The strategy for the proof of the lemma is now the following: We first define the extension ϕ on the cells of P_2 by linear interpolation between u and v. Then we define ϕ on the cells of $P_3, \ldots, P_m$ recursively by a homogeneous extension of degree 0 of $\phi \circ \Phi_p^{-1}$, in each step taking advantage of the fact that the map is already defined on the boundary of each given cell p. We now describe the construction in detail.

Suppose $q \in Q_1$ is given. For two points $x, y \in q$ we have

$$\begin{aligned}
&|u(x) - v(x)|^2 - |u(y) - v(y)|^2 \\
&\qquad \le \int_q |\nabla^q |u - v|^2| \, dh^1 \\
&\qquad = 2 \int_q |u - v||\nabla^q u - \nabla^q v| \, dh^1 \\
&\qquad \le 4 \left(\int_q |u - v|^2 \, dh^1 \right)^{1/2} \left(\int_q (|\nabla u|^2 + |\nabla v|^2) \, dh^1 \right)^{1/2}.
\end{aligned}$$

Since (4.9) and (4.10) hold, it follows that

$$\begin{aligned}
&\|u - v\|_{L^\infty(q)} \\
&\qquad \le C\lambda^{2-m} \left(\int_{S_1(0)} (|\nabla u|^2 + |\nabla v|^2) \, do \right)^{1/2} \left(\int_{S_1(0)} |u - v|^2 \, do \right)^{1/2} \\
&\qquad\quad + C\lambda^{1-m} \int_{S_1(0)} |u - v|^2 \, do.
\end{aligned}$$

Defining ϕ on $p = \bar{q}$ by

$$\phi(sx) = \frac{s - 1 + \lambda}{\lambda} u(x) + \frac{1 - s}{\lambda} v(x), \quad x \in q, \; 1 - \lambda \le s \le \lambda,$$

we therefore have

$$\begin{aligned}
&\operatorname{dist}^2(\phi(x), K) \\
&\quad \le C\lambda^{2-m} \left(\int_{S_1(0)} (|\nabla u|^2 + |\nabla v|^2) \, do \right)^{1/2} \left(\int_{S_1(0)} |u - v|^2 \, do \right)^{1/2} \\
&\qquad + C\lambda^{1-m} \int_{S_1(0)} |u - v|^2 \, do \qquad (4.12)
\end{aligned}$$

for h^2-almost every $x \in p$. Moreover,

$$\int_p |\nabla^p \phi|^2 \, dh^2 \le C\lambda \int_q (|\nabla u|^2 + |\nabla v|^2) \, dh^1 + \frac{C}{\lambda} \int_q |u - v|^2 \, dh^1.$$

If ϕ is defined on each $p \in P_2$ in this way, then (4.9) and (4.10) imply

$$\sum_{p \in P_2} \int_p |\nabla^p \phi|^2 \, dh^2 \le C\lambda^{3-m} \int_{S_1(0)} (|\nabla u|^2 + |\nabla v|^2) \, do + C\lambda^{1-m} \int_{S_1(0)} |u - v|^2 \, do. \tag{4.13}$$

Now suppose $p \in P_3$. On the boundary ∂p, the map ϕ is already determined by the previous construction and the conditions $\phi(x) = u(x)$ on $S_1(0)$ and $\phi(x) = v(x/(1-\lambda))$ on $S_{1-\lambda}(0)$. We extend ϕ into the interior of p by

$$\phi(x) = \phi\left(\Phi_p^{-1}\left(\lambda \frac{\Phi_p(x)}{|\Phi_p(x)|}\right)\right), \quad x \in p.$$

Then

$$\int_p |\nabla^p \phi|^2 \, dh^3 \le C\lambda \int_{\partial p} |\nabla^{\partial p} \phi|^2 \, dh^2.$$

Summing over P_3 and taking (4.9) and (4.13) into account, we see that

$$\sum_{p \in P_3} \int_p |\nabla^p \phi|^2 \, dh^3 \le C\lambda^{4-m} \int_{S_1(0)} (|\nabla u|^2 + |\nabla v|^2) \, do + C\lambda^{2-m} \int_{S_1(0)} |u - v|^2 \, do.$$

We use the same method to define ϕ on the cells of $P_4, \ldots, P_m$. Each time we obtain the inequality

$$\sum_{p \in P_j} \int_p |\nabla^p \phi|^2 \, dh^j \le C\lambda^{j+1-m} \int_{S_1(0)} (|\nabla u|^2 + |\nabla v|^2) \, do + C\lambda^{j-1-m} \int_{S_1(0)} |u - v|^2 \, do.$$

For $j = m$, this is inequality (4.7). Inequality (4.8) follows from (4.12) and the construction. □

Remark 4.3 *Lemma 4.4 is also useful for studying energy minimizing maps as described in Section 3.9.2, although we will use it for a different purpose.*

4.4 More About the Tangent Frame Field

Before we return to the problem of regularity of almost harmonic maps, we need to know a few more facts about the tangent frame field $(e_1, \ldots, e_k)$ constructed in Section 3.6. First we recall briefly the construction. Suppose $u \in H^1(\Omega; N)$. We assume that there exist k smooth vector fields $\bar{e}_1, \ldots, \bar{e}_k$, where k is the dimension of N, at least in a neighborhood of the closure of $u(\Omega)$, such that for any point y in this neighborhood, the vectors $\bar{e}_1(y), \ldots, \bar{e}_k(y)$ form an orthonormal basis of the tangent space T_yN. If such vector fields do not exist, we can replace N by a larger manifold by the same arguments as in Section 3.6, and then we can find vector fields with the required properties. Suppose now $B_r(x_0) \subset \Omega$. We consider the functional

$$I(R) = \frac{1}{2}\sum_{i=1}^{k}\int_{B_r(x_0)} |\nabla e_i|^2\,dx$$

for $R \in H^1(B_r(x_0); \mathrm{SO}(k))$, where

$$e_i = \sum_{j=1}^{k} R_{ij}(\bar{e}_j \circ u).$$

Let $R \in H^1(B_r(x_0); \mathrm{SO}(k))$ be a minimizer of I and

$$\omega_{ij}(x) = \begin{cases} \langle \nabla e_i(x), e_j(x)\rangle & \text{if } x \in B_r(x_0), \\ 0 & \text{if } x \notin B_r(x_0), \end{cases}$$

then we have the Euler-Lagrange equations

$$\operatorname{div}\omega_{ij} = 0 \quad \text{in } \mathbb{R}^m. \tag{4.14}$$

Moreover,

$$\int_{B_r(x_0)} |\omega|^2\,dx \le \int_{B_r(x_0)} |\nabla e|^2\,dx \le C\int_{B_r(x_0)} |\nabla u|^2\,dx, \tag{4.15}$$

where $e = (e_1, \dots, e_k)$ and $\omega = (\omega_{ij})_{1 \le i,j \le k}$. The purpose of the following arguments is to find better estimates for the energy of e and thus for the L^2-norm of ω.

Lemma 4.5 *There exists a number $\bar{\epsilon} > 0$, depending only on m and N, such that for every $\epsilon_0 \in (0, \bar{\epsilon}]$ there are constants $\epsilon > 0$ and $\theta \in (0,1)$, depending only on m, N, and ϵ_0, with the following property: Suppose $x_1 \in B_r(x_0)$ and $s \in (0, r]$ are such that*

$$s^{2-m} \int_{B_r(x_0) \cap B_s(x_1)} |\nabla u|^2 \, dx \le \epsilon$$

and

$$s^{2-m} \int_{B_r(x_0) \cap B_s(x_1)} |\nabla R|^2 \, dx \le \epsilon_0.$$

Then

$$(\theta s)^{2-m} \int_{B_r(x_0) \cap B_{\theta s}(x_1)} |\nabla R|^2 \, dx \le \epsilon_0.$$

Proof. Note that

$$I(R) = J(R) + \sum_{i=1}^{k} \int_{B_r(x_0)} |\nabla(\bar{e}_i \circ u)|^2 \, dx,$$

where

$$J(R) = \int_{B_r(x_0)} |\nabla R|^2 \, dx + 2 \sum_{i,j,l=1}^{k} \int_{B_r(x_0)} R_{ij} \nabla R_{jl} \cdot \langle \nabla(\bar{e}_j \circ u), \bar{e}_l \circ u \rangle \, dx,$$

because we have $R \in \mathrm{SO}(k)$ and $\langle \bar{e}_i \circ u, \bar{e}_j \circ u \rangle = \delta_{ij}$ almost everywhere. In particular R is a minimizer of the functional J. For any fixed ball $B_s(x_1) \subset B_r(x_0)$, R is also a minimizer of

$$\begin{aligned} J(R; u, B_s(x_1)) = & \int_{B_r(x_0) \cap B_s(x_1)} |\nabla R|^2 \, dx \\ & + 2 \sum_{i,j,l=1}^{k} \int_{B_r(x_0) \cap B_s(x_1)} R_{ij} \nabla R_{jl} \cdot \langle \nabla(\bar{e}_j \circ u), \bar{e}_l \circ u \rangle \, dx \end{aligned}$$

among all maps in $H^1(B_r(x_0) \cap B_s(x_1); \mathrm{SO}(k))$ that coincide with R on $B_r(x_0) \cap S_s(x_1)$. Note further that

$$\left| J(S; u, B_s(x_1)) - \int_{B_r(x_0)\cap B_s(x_1)} |\nabla S|^2 \, dx \right| \\ \leq C \left(\int_{B_r(x_0)\cap B_s(x_1)} |\nabla S|^2 \, dx \right)^{1/2} \left(\int_{B_r(x_0)\cap B_s(x_1)} |\nabla u|^2 \, dx \right)^{1/2} . \quad (4.16)$$

Suppose now the claim were false. We may assume that it fails for $B_s(x_1) = B_1(0)$, otherwise we consider the maps $u(sx + x_1)$ and $R(sx + x_1)$ instead of u and R. For the ease of notation we also assume that the ball is entirely contained in $B_r(x_0)$. Everything we do in this proof, however, can be done also if this assumption is not true.

Under these conditions, we can choose for any fixed number $\bar{\epsilon} > 0$ some $\epsilon_0 \in (0, \bar{\epsilon}]$, such that for any $\theta \in (0, \frac{1}{2})$ the following holds. There are maps $u_l \in H^1(B_1(0); N)$, $l \in \mathbb{N}$, and corresponding maps $R_l \in H^1(B_1(0); \mathrm{SO}(k))$ which minimize the functional $J(\,\cdot\,; u_l, B_1(0))$ for their boundary values, such that

$$\int_{B_1(0)} |\nabla u_l|^2 \, dx = \epsilon_l \to 0 \quad \text{as } l \to \infty \quad (4.17)$$

and

$$\int_{B_1(0)} |\nabla R_l|^2 \, dx \leq \epsilon_0,$$

but

$$\theta^{2-m} \int_{B_\theta(0)} |\nabla R_l|^2 \, dx > \epsilon_0.$$

There exist radii $r_l \in [\frac{1}{2}, 1]$ such that

$$\int_{S_{r_l}(0)} |\nabla R_l|^2 \, do \leq C\epsilon_0.$$

We rescale the maps and replace θ and ϵ_0 by larger constants, then we may assume that

$$\int_{B_1(0)} |\nabla R_l|^2 \, dx \leq \epsilon_0, \quad (4.18)$$

$$\int_{S_1(0)} |\nabla R_l|^2 \, do \leq \epsilon_0, \tag{4.19}$$

and

$$\theta^{2-m} \int_{B_\theta(0)} |\nabla R_l|^2 \, dx > \frac{\epsilon_0}{C_0}, \tag{4.20}$$

where C_0 is a universal constant. Moreover, the condition (4.17) still holds (with different values for ϵ_l). In particular the maps R_l are uniformly bounded in $H^1(B_1(0); \mathbb{R}^{k\times k})$, and their traces on $S_1(0)$ are uniformly bounded in $H^1(S_1(0); \mathbb{R}^{k\times k})$. We may thus assume that they converge weakly in $H^1(B_1(0); \mathbb{R}^{k\times k})$ to a limit $R_0 \in H^1(B_1(0); \mathrm{SO}(k))$, and that the traces on $S_1(0)$ converge weakly in $H^1(S_1(0); \mathbb{R}^{k\times k})$ as well as strongly in $L^2(S_1(0); \mathbb{R}^{k\times k})$ to the corresponding trace of R_0. Otherwise we choose a subsequence with these properties. Obviously we have

$$\int_{S_1(0)} |\nabla R_0|^2 \, do \leq \epsilon_0.$$

We want to prove that R_0 is a minimizer of the Dirichlet energy in $B_1(0)$ among all maps with the same boundary values. To this end, we choose a map $S \in H^1(B_1(0); \mathrm{SO}(k))$ that coincides with R_0 on $S_1(0)$ in the sense of traces. Fix $\lambda \in (0, \frac{1}{2}]$. According to Lemma 4.4, we can find maps $T_l \in H^1(B_1(0)\backslash B_{1-\lambda}(0); \mathbb{R}^{k\times k})$ which agree with R_l on $S_1(0)$ and with $R_0(x/(1-\lambda))$ on $S_{1-\lambda}(0)$, such that

$$\int_{B_1(0)\backslash B_{1-\lambda}(0)} |\nabla T_l|^2 \, dx \leq C\lambda\epsilon_0 + \frac{C}{\lambda} \int_{S_1(0)} |R_l - R_0|^2 \, do \tag{4.21}$$

and

$$\begin{aligned}\operatorname{dist}^2(T_l(x), \mathrm{SO}(k)) \leq{}& C\lambda^{2-m}\sqrt{\epsilon_0} \left(\int_{S_1(0)} |R_l - R_0|^2 \, do \right)^{1/2} \\ &+ C\lambda^{1-m} \int_{S_1(0)} |R_l - R_0|^2 \, do\end{aligned}$$

for almost every $x \in B_1(0)\backslash B_{1-\lambda}(0)$. In particular T_l takes on values close to $\mathrm{SO}(k)$ almost everywhere, whenever l is sufficiently large. Thus T_l can be composed with the nearest point projection $\pi_{\mathrm{SO}(k)}$. The maps $\pi_{\mathrm{SO}(k)} \circ T_l$ also satisfy (4.21), but possibly for a larger constant C. We extend $\pi_{\mathrm{SO}(k)} \circ T_l$ by $S(x/(1-\lambda))$ in $B_{1-\lambda}(0)$ and denote the resulting map also by T_l for simplicity. Then we have

$$\int_{B_1(0)} |\nabla R_l|^2\, dx \le J(R_l; u_l, B_1(0)) + C\sqrt{\epsilon_0\epsilon_l}$$
$$\le J(T_l; u_l, B_1(0)) + C\sqrt{\epsilon_0\epsilon_l}$$

by (4.16) and the minimizing property of R_l. Hence

$$\begin{aligned}\int_{B_1(0)} |\nabla R_l|^2\, dx &\le \int_{B_1(0)} |\nabla T_l|^2\, dx + C\sqrt{\epsilon_l}\left(\epsilon_0 + \int_{B_1(0)} |\nabla T_l|^2\, dx\right)^{1/2} \\ &\le (1-\lambda)^{m-2}\int_{B_1(0)} |\nabla S|^2\, dx + C\lambda\epsilon_0 \\ &\quad + \frac{C}{\lambda}\int_{S_1(0)} |R_l - R_0|^2\, do \\ &\quad + C\sqrt{\epsilon_l}\left(\epsilon_0 + \int_{B_1(0)} |\nabla T_l|^2\, dx\right)^{1/2}.\end{aligned}$$

Since λ has been chosen arbitrarily, and since

$$\int_{S_1(0)} |R_l - R_0|^2\, do$$

and ϵ_l both converge to 0 as $l \to \infty$, we find that

$$\begin{aligned}\int_{B_1(0)} |\nabla R_0|^2\, dx &\le \liminf_{l\to\infty} \int_{B_1(0)} |\nabla R_l|^2\, dx \\ &\le \limsup_{l\to\infty} \int_{B_1(0)} |\nabla R_l|^2\, dx \le \int_{B_1(0)} |\nabla S|^2\, dx.\end{aligned}$$

That is, R_0 indeed minimizes the Dirichlet energy. In particular it is a stationary weakly harmonic map into $\mathrm{SO}(k)$. Moreover, since the last inequality is true in particular for $S = R_0$, we have strong convergence of R_l to R_0 in $H^1(B_1(0); \mathbb{R}^{k\times k})$, not only weak convergence. The results of Section 3.7 imply that whenever ϵ_0 and θ are small enough, then

$$\theta^{2-m}\int_{B_\theta(0)} |\nabla R_0|^2\, dx \le \frac{\epsilon_0}{2C_0}$$

for the constant C_0 of (4.20). But this together with the strong convergence establishes a contradiction to (4.20). This concludes the proof. □

Lemma 4.6 *For any $\epsilon_0 > 0$ there exists some $\epsilon > 0$, depending only on m, N, and ϵ_0, such that*

$$\sup_{x_1 \in B_r(x_0)} \sup_{0<s\leq r} \left(s^{2-m} \int_{B_r(x_0)\cap B_s(x_1)} |\nabla e|^2 \, dx \right) \leq \epsilon_0 \tag{4.22}$$

whenever

$$\sup_{x_1 \in B_r(x_0)} \sup_{0<s\leq r} \left(s^{2-m} \int_{B_r(x_0)\cap B_s(x_1)} |\nabla u|^2 \, dx \right) \leq \epsilon.$$

Proof. To prove (4.22), it suffices to use (4.15) and then apply Lemma 4.5 repeatedly. □

In particular we obtain good estimates for the BMO-seminorm of e if we combine Lemma 4.6 with the Poincaré inequality. Next we want to exploit Lemma 4.6 in order to estimate the L^2-norm of Ω, using the observation that

$$|\omega_{ij}|^2 = \sum_{i,j=1}^{k} \langle \nabla e_i, e_j \rangle \cdot \omega_{ij}.$$

The right-hand side has the structure that is needed to apply Theorem 2.3 and Theorem 2.4. Hence if we extend e to $\mathbb{R}^m$ such that the L^2-norm of the gradient and the BMO-seminorm are increased at most by some fixed constant, then Lemma 4.6, Theorem 2.3, and Theorem 2.4 imply

$$\|\omega\|^2_{L^2(\mathbb{R}^m)} \leq C\sqrt{\epsilon_0}\|\nabla u\|^2_{L^2(B_r(x_0))}.$$

This inequality, however, is not quite good enough yet for our purpose. With a localized version of the arguments in the proof of Theorem 2.4, we are able to prove a better estimate.

Lemma 4.7 *Suppose $\omega \in L^2(B_1(0);\mathbb{R}^m)$ satisfies $\operatorname{div}\omega = 0$ in $B_1(0)$ in the distribution sense. Suppose further that $\tilde{\omega} = \omega|_{S_1(0)} \in L^2(S_1(0);\mathbb{R}^m)$, i. e.*

$$\int_{B_1(0)} \nabla\eta \cdot \omega \, dx = \int_{S_1(0)} \eta \, x \cdot \tilde{\omega} \, do$$

for all $\eta \in C^\infty(\mathbb{R}^m)$. Let $f \in H^1(B_1(0))$ be a function with the trace $\tilde{f} = f|_{S_1(0)} \in H^1(S_1(0))$. Then there exists a number $a \in \mathbb{R}$, such that the *function*

$$g(x) = \begin{cases} \nabla f(x) \cdot \omega(x) - a & \text{if } x \in B_1(0), \\ 0 & \text{if } x \notin B_1(0) \end{cases}$$

is in $\mathcal{H}^1(\mathbb{R}^m)$ *with*

$$\|g\|_{\mathcal{H}^1(\mathbb{R}^m)} \\ \leq C\left(\|\nabla f\|_{L^2(B_1(0))} + \|\nabla^{S_1(0)}\tilde{f}\|_{L^2(S_1(0))}\right)\left(\|\omega\|_{L^2(B_1(0))} + \|\tilde{\omega}\|_{L^2(S_1(0))}\right)$$

for a constant C *that depends only on* m*. Here* $\nabla^{S_1(0)}$ *denotes the gradient on the manifold* $S_1(0)$*. Moreover,*

$$|a| \leq C\left(\inf_{\alpha\in\mathbb{R}} \|\tilde{f} - \alpha\|_{L^2(S_1(0))}\right)\|\tilde{\omega}\|_{L^2(S_1(0))}.$$

Proof. Set

$$a = \fint_{B_1(0)} \nabla f \cdot \omega \, dx.$$

Then we also have

$$a = \frac{1}{\omega_m}\int_{S_1(0)} (\tilde{f} - \alpha)\, x \cdot \tilde{\omega}\, do$$

for any $\alpha \in \mathbb{R}$, and the inequality for a follows. Moreover,

$$\int_{B_1(0)} g\, dx = 0.$$

Let $\phi \in C_0^\infty(B_1(0))$ with

$$\sup_{B_1(0)} |\nabla\phi| \leq 1.$$

That is, ϕ is in the set $\mathcal{F}$ defined in Section 2.4. For $s > 0$, set $\phi_s(x) = s^{-m}\phi(x/s)$. Then for every $x \in \mathbb{R}^m$ and every $s \geq \frac{1}{4}$,

$$\begin{aligned} |(\phi_s * g)(x)| &= \left|\int_{B_1(0)} (\phi_s(x-y) - \phi_s(x))g(y)\, dy\right| \\ &\leq \|\nabla\phi_s\|_{L^\infty(\mathbb{R}^m)} \int_{B_1(0)} |g|\, dy \\ &\leq Cs^{-m-1}\left(\|\nabla f\|_{L^2(B_1(0))}\|\omega\|_{L^2(B_1(0))} + |a|\right). \end{aligned}$$

If $x \notin B_{1+s}(0)$, then we even have $(\phi_s * g)(x) = 0$. Hence

$$\sup_{s\geq 1/4} |(\phi_s * g)(x)| \leq C(|x|+1)^{-m-1}\left(\|\nabla f\|_{L^2(B_1(0))}\|\omega\|_{L^2(B_1(0))} + |a|\right).$$

Thus it suffices to show that the inequality

$$\int_{\mathbb{R}^m} \sup_{0<s\leq\frac{1}{4}} |\phi_s * (\nabla f\cdot\omega)|\,dx$$
$$\leq C\left(\|\nabla f\|_{L^2(B_1(0))} + \|\nabla^{S_1(0)}\tilde{f}\|_{L^2(S_1(0))}\right)\left(\|\omega\|_{L^2(B_1(0))} + \|\tilde{\omega}\|_{L^2(S_1(0))}\right)$$

holds. Here and in the rest of the proof we extend ∇f and ω by 0 outside of $B_1(0)$ in order to simplify the notation.

Let $x \in \mathbb{R}^m$ and $0 < s \leq \frac{1}{4}$. We have

$$\begin{aligned}(\phi_s * (\nabla f\cdot\omega))(x) &= \int_{B_1(0)} \phi_s(x-y)\nabla f(y)\cdot\omega(y)\,dy \\ &= \int_{S_1(0)\cap B_s(x)} \phi_s(x-y)(\tilde{f}(y) - \bar{f}_{B_s(x)})\,x\cdot\tilde{\omega}(y)\,do(y) \\ &\quad + \int_{B_s(x)} (f(y) - \bar{f}_{B_s(x)})\nabla\phi_s(x-y)\cdot\omega(y)\,dy.\end{aligned}$$

We estimate

$$\begin{aligned}&\int_{B_s(x)} (f(y) - \bar{f}_{B_s(x)})\nabla\phi_s(x-y)\cdot\omega(y)\,dy \\ &\leq Cs^{-m-1}\left(\int_{B_s(x)} |f-\bar{f}_{B_s(x)}|^{\frac{2m}{m-1}}\,dy\right)^{\frac{m-1}{2m}}\left(\int_{B_s(x)}|\omega|^{\frac{2m}{m+1}}\,dy\right)^{\frac{m+1}{2m}} \\ &\leq C\left(\fint_{B_s(x)} |\nabla f|^{\frac{2m}{m+1}}\,dy\right)^{\frac{m+1}{2m}}\left(\fint_{B_s(x)}|\omega|^{\frac{2m}{m+1}}\,dy\right)^{\frac{m+1}{2m}}\end{aligned}$$

with the Hölder and the Sobolev inequality.

We extend $\tilde{f}$ and $\tilde{\omega}$ to $\mathbb{R}^m$ by

$$\tilde{f} = \begin{cases}\tilde{f}(x/|x|) & \text{if } x \in \overline{B_1(0)}\backslash\{0\}, \\ 0 & \text{else,}\end{cases}$$

and

$$\tilde{\omega} = \begin{cases}\tilde{\omega}(x/|x|) & \text{if } x \in \overline{B_1(0)}\backslash\{0\}, \\ 0 & \text{else.}\end{cases}$$

If $B_s(x) \cap S_1(0) \neq \emptyset$, let $x_1 \in S_1(0)$ be the point satisfying $|x - x_1| = 1 - |x|$. Then

$$\int_{S_1(0)\cap B_s(x)} \phi_s(x-y)(\tilde{f}(y) - \bar{f}_{B_s(x)})\, x \cdot \tilde{\omega}(y)\, do(y)$$
$$\leq C s^{-m-1} \int_{B_{2s}(x_1)} (|\tilde{f} - f| + |f - \bar{f}_{B_s(x)}|)|\tilde{\omega}|\, dy.$$

Now similar estimates as above can be obtained. The Sobolev inequality can be applied to $\tilde{f} - f$ in $B_{2s}(x_1)$, because $\tilde{f} - f = 0$ on $S_1(0)$. The claim of the lemma then follows from the maximal theorem, with the same arguments as in the proof of Theorem 2.4. □

Lemma 4.8 *Suppose (4.22) holds for some $\epsilon_0 \leq 1$. Let $B_s(x_1) \subset B_r(x_0)$. Then for any $\lambda > 0$, there exist two constants $\theta \in (0,1)$ and $C_0 > 0$, depending only on m, N, and λ, such that*

$$(\theta s)^{2-m} \int_{B_{\theta s}(x_1)} |\omega|^2\, dx \leq C_0 \sqrt{\epsilon_0} \left([e]^2_{\mathrm{BMO}(B_s(x_1))} + s^{2-m} \int_{B_s(x_0)} |\omega|^2\, dx \right)$$
$$+ \lambda \left(s^{1-m} \int_{B_s(x_1)} |\nabla e|\, dx \right)^2.$$

Proof. We assume for convenience that $B_r(x_0) = B_1(0)$. For any other ball, we can rescale all the involved maps accordingly. Choose a number $\rho \in [\frac{1}{2}, \frac{3}{4}]$ such that

$$\int_{S_\rho(0)} |\omega|^2\, do \leq C \int_{B_1(0)} |\omega|^2\, dx,$$

$$\int_{S_\rho(0)} |\nabla e|^2\, do \leq C \int_{B_1(0)} |\nabla e|^2\, dx,$$

and

$$\int_{S_\rho(0)} |e - \bar{e}_{B_1(0)}|^q\, do \leq C \int_{B_1(0)} |e - \bar{e}_{B_1(0)}|^q\, dx \leq C_1 [e]^q_{\mathrm{BMO}(B_1(0))}$$

for a number $q \in (m, \infty)$ which we will determine later. The last inequality follows from Lemma 2.5. Here C_1 is a number that depends only on m and q. Let v_{ij} be the solutions of

$$\begin{cases} \Delta v_{ij} = 0 & \text{in } B_\rho(0), \\ v_{ij} = \langle e_i - \bar{e}_i, e_j \rangle & \text{on } S_\rho(0), \end{cases}$$

where

$$\bar{e}_i = \fint_{B_1(0)} e_i \, dx.$$

The traces $v_{ij}|_{S_\rho(0)}$ are in the Sobolev space $W^{1,2q/(q+2)}(S_\rho(0))$ and satisfy the inequality

$$\|\nabla^{S_\rho(0)} v_{ij}\|_{L^{2q/(q+2)}(S_\rho(0))} \leq C_2 \left(\|\omega\|_{L^2(B_1(0))} + [e]_{\mathrm{BMO}(B_1(0))}\|\nabla e\|_{L^2(B_1(0))}\right)$$

for a constant C_2 that depends only on m and q. Here $\nabla^{S_\rho(0)}$ denotes the gradient on the sphere $S_\rho(0)$.

In the following we use certain results from the theory of elliptic boundary value problems that we cannot discuss in this book. The reader who is unfamiliar with this theory, can find the results, for instance, in a book by J.-L. Lions and E. Magenes [1972]. We first observe that the space $W^{1,2q/(q+2)}(S_\rho(0))$ is embedded continuously into $H^\sigma(S_\rho(0))$, where $\sigma = 1 - (m-1)/q$. The theory of boundary value problems then implies $v_{ij} \in H^{\sigma+1/2}(B_\rho(0))$, and we have a corresponding estimate for the norm. This space is embedded in $W^{1,\gamma}(B_\rho(0))$, where

$$\frac{1}{\gamma} = \frac{1}{2} - \frac{\sigma - 1/2}{m},$$

and the last quantity is strictly less than $\frac{1}{2}$, provided that q is chosen sufficiently large. Thus we have

$$\|\nabla v_{ij}\|_{L^\gamma(B_\rho(0))} \leq C_3 \left(\|\omega\|_{L^2(B_1(0))} + \sqrt{\epsilon_0}[e]_{\mathrm{BMO}(B_1(0))}\right)$$

for a constant C_3 that depends only on m and q.

Now we note that

$$\begin{aligned}\int_{B_\rho(0)} |\omega_{ij} - \nabla v_{ij}|^2 \, dx &= \int_{B_\rho(0)} (\omega_{ij} - \nabla v_{ij}) \cdot (\langle \nabla(e_i - \bar{e}_i), e_j\rangle - \nabla v_{ij}) \, dx \\ &= -\int_{B_\rho(0)} (\omega_{ij} - \nabla v_{ij}) \cdot \langle e_i - \bar{e}_i, \nabla e_j\rangle \, dx. \qquad (4.23)\end{aligned}$$

We have

$$\begin{aligned}&\int_{B_\rho(0)} \nabla v_{ij} \cdot \langle e_i - \bar{e}_i, \nabla e_j\rangle \, dx \\ &\qquad \leq \|\nabla e\|_{L^2(B_1(0))}\|\nabla v_{ij}\|_{L^\gamma(B_\rho(0))}\|e - \bar{e}_{B_1(0)}\|_{L^{2\gamma/(\gamma-2)}(B_\rho(0))} \\ &\qquad \leq C_4 \left(\sqrt{\epsilon_0}\|\omega\|_{L^2(B_1(0))} + \epsilon_0[e]_{\mathrm{BMO}(B_1(0))}\right)[e]_{\mathrm{BMO}(B_1(0))} \qquad (4.24)\end{aligned}$$

for a constant C_4 that depends only on m and q. Moreover,

$$-\int_{B_\rho(0)} \omega_{ij} \cdot \langle e_i - \bar{e}_i, \nabla e_j \rangle \, dx \leq C\sqrt{\epsilon_0} \|\omega\|_{L^2(B_1(0))} [e]_{\mathrm{BMO}(B_1(0))}. \quad (4.25)$$

The last estimate is a consequence of Lemma 4.7 and Theorem 2.3. Finally, for $\theta \leq \rho$ we have

$$\begin{aligned} \theta^{2-m} \int_{B_\theta(0)} |\nabla v_{ij}|^2 \, dx &\leq C\theta^2 \left(\int_{B_\rho(0)} |\nabla v_{ij}| \, dx \right)^2 \\ &\leq C\theta^2 \int_{B_\rho(0)} |\omega_{ij} - \nabla v_{ij}|^2 \, dx \\ &\quad + C\theta^2 \left(\int_{B_1(0)} |\nabla e| \, dx \right)^2 . \end{aligned} \quad (4.26)$$

To finish the proof, we now merely have to combine (4.23) with the estimates (4.24), (4.25), and (4.26). □

4.5 An Energy Inequality

For $p = m/2$, the proofs of Section 4.2 break down. This is no surprise, as we cannot expect Hölder continuity for solutions of (4.1) in this case. Even solutions of the linear equation

$$\Delta u = f$$

may be discontinuous if we only have $f \in L^{m/2}(\Omega)$. Nevertheless, we can still prove an inequality for the energy of an almost harmonic map that will turn out to be useful for some other results in this and the next chapter. Before we state and prove this result, we derive a few preliminary estimates.

We assume in the following that $u \in H^1(\Omega; N)$ is a solution of (4.1) for a function $f \in L^p(\Omega; \mathbb{R}^n)$, where $p = 2$ if $m \leq 4$ and $p = m/2$ else. We consider a ball $B_r(x_0) \subset \Omega$. To simplify the presentation, we assume that $B_r(x_0) = B_1(0)$. Otherwise, we can consider the map $u(rx + x_0)$ instead of u and thus transform $B_r(x_0)$ into $B_1(0)$. We assume that

$$\sup_{x_0 \in B_1(0)} \sup_{0 < r \leq 1} \left(r^{2-m} \int_{B_1(0) \cap B_r(x_0)} |\nabla u|^2 \, dx \right) \leq \epsilon$$

for a number $\epsilon > 0$ that will be determined later. We use again the tangent frame field $e = (e_1, \ldots, e_k)$ constructed in Section 3.6. We note that

$$\operatorname{div} \langle \nabla u, e_i \rangle = \sum_{j=1}^{k} \langle \nabla u, e_j \rangle \cdot \omega_{ij} + \langle f, e_i \rangle \tag{4.27}$$

for $i = 1, \ldots, k$. In the following arguments, we combine the properties of e derived in the previous section with the estimate from Lemma 4.3.

First we fix two numbers $\epsilon_0 > 0$ and $\lambda \in (0, 1]$. If ϵ is sufficiently small, we have

$$\sup_{x_0 \in B_1(0)} \sup_{0<r\leq 1} \left(r^{2-m} \int_{B_1(0)\cap B_r(x_0)} |\nabla e|^2 \, dx \right) \leq \epsilon_0$$

by Lemma 4.6. For any ball $B_{2r}(x_0) \subset B_1(0)$ and for $q \geq 1$ and $F \in L^q(B_{2r}(x_0))$, we define

$$\Psi^q(F; x_0, r) = \sup_{x_1 \in B_r(x_0)} \sup_{0<s\leq r} \left(s^{q-m} \int_{B_s(x_1)} |F|^q \, dx \right)^{1/q}.$$

Let $\theta \in (0, \frac{1}{2})$ be a number such that the inequality of Lemma 4.8 is satisfied for the given λ. We have

$$|\nabla e_i| \leq C \left(\sum_{j=1}^{k} |\omega_{ij}| + |\nabla u| \right),$$

thus

$$\Psi^1(|\nabla e|; x_0, r) \leq C \sup_{x_1 \in B_r(x_0)} \sup_{0<s\leq \theta r} \left(s^{1-m} \int_{B_s(x_1)} (|\omega| + |\nabla u|) \, dx \right) + (\theta r)^{1-m} \int_{B_{2r}(x_0)} |\nabla e| \, dx.$$

By Lemma 4.8 and the Poincaré inequality,

$$\sup_{x_1 \in B_r(x_0)} \sup_{0<s\leq \theta r} \left(s^{1-m} \int_{B_s(x_1)} |\omega| \, dx \right) \leq (C_1 \epsilon_0^{1/4} + \sqrt{\lambda}) \Psi^1(|\nabla e|; x_0, 2r) + C_1 \epsilon_0^{1/4} \Psi^2(|\omega|; x_0, r)$$

for a constant C_1 that depends only on m, N, and λ, provided that $B_{4r}(x_0) \subset B_1(0)$. Similarly, by Lemma 4.3, we have

$$\sup_{x_1 \in B_r(x_0)} \sup_{0<s\leq \theta r} \left(s^{1-m} \int_{B_s(x_1)} |\nabla u|\, dx \right) \\ \leq C(\sqrt{\epsilon} + \theta)\Psi^1(|\nabla u|; x_0, 2r) + C\|f\|_{L^{m/2}(B_{2r}(x_0))}.$$

Hence for any given $\gamma > 0$, we can choose ϵ, ϵ_0, λ, and θ so small that

$$\Psi^1(|\nabla e|; x_0, r) \leq \frac{\gamma}{2}\left(\Psi^1(|\nabla e|; x_0, 2r) + \Psi^1(|\nabla u|; x_0, 2r) + \Psi^2(|\omega|; x_0, r)\right) \\ + (\theta r)^{1-m} \int_{B_{2r}(x_0)} |\nabla e|\, dx + C\|f\|_{L^{m/2}(B_{2r}(x_0))}.$$

Similarly we conclude that

$$\begin{aligned} \Psi^2(|\omega|; x_0, r) &\leq C_2\epsilon_0^{1/4}(\Psi^1(|\nabla e|; x_0, 2r/\theta) + \Psi^2(|\omega|; x_0, r/\theta)) \\ &\quad + \sqrt{\lambda} \sup_{x_1 \in B_r(x_0)} \sup_{0 \leq s \leq r} \left(s^{1-m} \int_{B_s(x_1)} |\nabla e|\, dx \right) \\ &\quad + \sqrt{\lambda} \sup_{x_1 \in B_r(x_0)} \sup_{r \leq s \leq r/\theta} \left(s^{1-m} \int_{B_s(x_1)} |\nabla e|\, dx \right) \\ &\leq C_2\epsilon_0^{1/4}\left(\Psi^1(|\nabla e|; x_0, 2r/\theta) + \Psi^2(|\omega|; x_0, r/\theta)\right) \\ &\quad + \sqrt{\lambda}\Psi^1(|\nabla e|; x_0, r) + \sqrt{\lambda} r^{1-m} \int_{B_{2r/\theta}(x_0)} |\nabla e|\, dx, \end{aligned}$$

using Lemma 4.8, where C_2 is a constant that depends only on m, N, and λ. Here we assume that $B_{4r/\theta}(x_0) \subset B_1(0)$. Finally, by Lemma 4.3 and similar arguments,

$$\Psi^1(|\nabla u|; x_0, r) \leq \frac{\gamma}{2}\Psi^1(|\nabla u|; x_0, 2r) + (\theta r)^{1-m} \int_{B_{2r}(x_0)} |\nabla u|\, dx \\ + C\|f\|_{L^{m/2}(B_{2r}(x_0))},$$

if ϵ and θ are small enough.

Define now

$$X(x_0, r) = \Psi^1(|\nabla e|; x_0, r) + \Psi^1(|\nabla u|; x_0, r) + \Psi^2(|\omega|; x_0, r)$$

and

$$Y(x_0, r) = r^{1-m} \int_{B_r(x_0)} (|\nabla u| + |\nabla e|)\, dx.$$

Combining the previous estimates, we find

$$X(x_0, r) \le (\gamma + \sqrt{\lambda})X(x_0, 2r) + C_2\epsilon_0^{1/4} X(x_0, 2r/\theta) + C\theta^{1-m}(Y(x_0, 2r) + Y(x_0, 2r/\theta)) + C\|f\|_{L^{m/2}(B_{2r}(x_0))}.$$

Suppose $x_0 \in B_{1/4}(0)$. For $0 \le \delta \le \theta/4$, define

$$Z_\delta(x_0) = \sup_{\delta < r \le 1/4} \left(r^{-1} X(x_0, r)\right).$$

If $\delta > 0$, we have

$$\begin{aligned} Z_\delta(x_0) &\le \sup_{\delta < r \le \theta/8} \left(r^{-1} X(x_0, r)\right) + 8\theta^{-1} X(x_0, 1/4) \\ &\le (2\gamma + 2\sqrt{\lambda} + 2C_2\theta^{-1}\epsilon_0^{1/4}) Z_\delta(x_0) + C\theta^{-1}\sqrt{\epsilon + \epsilon_0} \\ &\quad + C\theta^{-m} \sup_{r>0} \fint_{B_1(0)\cap B_r(x_0)} (|\nabla u| + |\nabla e|)\, dx \\ &\quad + C \sup_{r>0} \left(r^{-1} \|f\|_{L^{m/2}(B_1(0)\cap B_r(x_0))}\right). \end{aligned}$$

If ϵ_0, γ, and λ are sufficiently small, we conclude that

$$Z_\delta(x_0) \le C\left(\sqrt{\epsilon + \epsilon_0} + F(x_0) + G(x_0)\right)$$

for a constant C that depends only on m and N, where

$$F(x_0) = \sup_{r>0} \fint_{B_1(0)\cap B_r(x_0)} (|\nabla u| + |\nabla e|)\, dx$$

and

$$G(x_0) = \sup_{r>0} (r^{-1} \|f\|_{L^{m/2}(B_1(0)\cap B_r(x_0))}).$$

Letting δ converge to 0, we obtain the same inequality for Z_0 instead of Z_δ. By the Hardy-Littlewood maximal theorem, we have

$$\|F\|_{L^2(B_1(0))} \le C\sqrt{\epsilon + \epsilon_0}.$$

Moreover,

$$G(x_0) \le C\|f\|^{1/2}_{L^{m/2}(B_1(0))} \sup_{r>0} \left(\fint_{B_1(0)\cap B_r(x_0)} |f|^{m/2}\, dx\right)^{1/m},$$

and it follows that

$$\|G\|_{L^2(B_1(0))} \le C\|f\|_{L^{m/2}(B_1(0))}.$$

Hence we have

$$\|Z_0\|_{L^2(B_{1/4}(0))} \leq C\left(\sqrt{\epsilon+\epsilon_0}+\|f\|_{L^{m/2}(B_1(0))}\right). \tag{4.28}$$

The same procedure can be carried out for $\Psi^1(|\nabla u|; x_0, r)$ alone. We omit the details here, for they are in fact less involved than the arguments above. We simply state the result: Setting

$$\tilde{Z}(x_0) = \sup_{0<r\leq 1/2}\left(r^{-1}\Psi^1(|\nabla u|; x_0, r)\right),$$

we find

$$\|\tilde{Z}\|_{L^2(B_{1/4}(0))} \leq C\left(\sqrt{\epsilon}+\|f\|_{L^{m/2}(B_1(0))}\right).$$

Thus for

$$\psi(x_0) = \sup_{0<r\leq 1/4}\left(r^{-1}[u]_{\mathrm{BMO}(B_r(x_0))}\right),$$

we have

$$\|\psi\|_{L^2(B_{1/4}(0))} \leq C\left(\sqrt{\epsilon}+\|f\|_{L^{m/2}(B_1(0))}\right) \tag{4.29}$$

by the Poincaré inequality. We will use the inequalities (4.28) and (4.29) in the next lemma.

Lemma 4.9 *Let $\zeta \in C_0^\infty(B_{1/4}(0))$. There exists a constant C_0, depending only on m, N, and ζ, with the following property. Set*

$$a_i = \fint_{B_1(0)} \zeta \nabla e_i \cdot \langle \nabla u, e_i\rangle\, dx.$$

Then the functions

$$g_i(x) = \begin{cases} \zeta(x)\nabla e_i(x)\cdot\langle\nabla u(x), e_i(x)\rangle - a_i & \text{if } x\in B_1(0),\\ 0 & \text{if } x \notin B_1(0),\end{cases}$$

are in $\mathcal{H}^1(\mathbb{R}^m;\mathbb{R}^n)$ and satisfy

$$\|g_i\|_{\mathcal{H}^1(\mathbb{R}^m)} \leq C_0\left(\sqrt{\epsilon}+\|f\|_{L^{m/2}(B_1(0))}\right)\left(\sqrt{\epsilon+\epsilon_0}+\|f\|_{L^{m/2}(B_1(0))}\right),$$

provided that ϵ and ϵ_0 are sufficiently small.

Proof. We proceed similarly as in the proof of Lemma 4.7. Let $\phi \in C_0^\infty(B_1(0))$ be a function satisfying

$$\sup_{B_1(0)} |\nabla\phi| \leq 1.$$

For $s > 0$, set $\phi_s(x) = s^{-m}\phi(x/s)$. Then by (4.27),

$$\begin{aligned}
&(\phi_s * (\zeta \nabla e_i \cdot \langle \nabla u, e_i \rangle))(x) \\
&= \int_{B_s(x)} \phi_s(x-y)\zeta(y)\nabla e_i(y) \cdot \langle \nabla u(y), e_i(y) \rangle \, dy \\
&= -\sum_{j=1}^{k} \int_{B_s(x)} \phi_s(x-y)\zeta(y) \left(e_i(y) - \overline{(e_i)}_{B_s(x)} \right) \langle \nabla u(y), e_j(y) \rangle \cdot \omega_{ij}(y) \, dy \\
&\quad - \int_{B_s(x)} \phi_s(x-y)\zeta(y) \left(e_i(y) - \overline{(e_i)}_{B_s(x)} \right) \langle f(y), e_i(y) \rangle \, dy \\
&\quad - \int_{B_s(x)} \phi_s(x-y) \left(e_i(y) - \overline{(e_i)}_{B_s(x)} \right) \nabla \zeta(y) \cdot \langle \nabla u(y), e_i(y) \rangle \, dy \\
&\quad + \int_{B_s(x)} \zeta(y) \left(e_i(y) - \overline{(e_i)}_{B_s(x)} \right) \nabla \phi_s(x-y) \cdot \langle \nabla u(y), e_i(y) \rangle \, dy.
\end{aligned}$$

The last term can be estimated by

$$C_1 \left(\fint_{B_s(x)} |\nabla e|^{\frac{2m}{m+1}} \, dx \right)^{\frac{m+1}{2m}} \left(\fint_{B_s(x)} |\nabla u|^{\frac{2m}{m+1}} \, dx \right)^{\frac{m+1}{2m}}$$

for a constant C_1 that depends only on m, N, and ζ, similarly as in the proof of Lemma 4.7. If $s \leq 1$, we also have

$$\begin{aligned}
&- \int_{B_s(x)} \phi_s(x-y) \left(e_i(y) - \overline{(e_i)}_{B_s(x)} \right) \nabla \zeta(y) \cdot \langle \nabla u(y), e_i(y) \rangle \, dy \\
&\qquad \leq C_2 \left(\fint_{B_s(x)} |\nabla e|^{\frac{2m}{m+1}} \, dx \right)^{\frac{m+1}{2m}} \left(\fint_{B_s(x)} |\nabla u|^{\frac{2m}{m+1}} \, dx \right)^{\frac{m+1}{2m}}
\end{aligned}$$

for another constant C_2 that depends only on m, N, and ζ. Moreover,

$$\begin{aligned}
&- \int_{B_s(x)} \phi_s(x-y)\zeta(y) \left(e_i(y) - \overline{(e_i)}_{B_s(x)} \right) \langle f(y), e_i(y) \rangle \, dy \\
&\qquad \leq C_3 \left(\fint_{B_s(x)} |\nabla e|^{\frac{2m}{m+1}} \, dx \right)^{\frac{m+1}{2m}} \left(\fint_{B_s(x)} |f|^{\frac{2m}{m+1}} \, dx \right)^{\frac{m+1}{2m}},
\end{aligned}$$

where C_3 depends only on m, N, and ζ. To estimate the terms in the sum, we can integrate by parts once more. We use the abbreviation

$$h_i^l = e_i^l - \overline{(e_i^l)}_{B_s(x)},$$

where $e_i = (e_i^1, \dots, e_i^n)$. Then we have

$$-\int_{B_s(x)} \phi_s(x-y)\zeta(y)h_i^l(y)\left\langle \nabla u(y), e_j(y)\right\rangle \cdot \omega_{ij}(y)\,dy$$

$$= \int_{B_s(x)} \phi_s(x-y)\left\langle u(y) - \bar{u}_{B_s(x)}, \nabla(\zeta(y)h_i^l(y)e_j(y))\right\rangle \cdot \omega_{ij}(y)\,dy$$

$$- \int_{B_s(x)} \zeta(y)h_i^l\left\langle u(y) - \bar{u}_{B_s(x)}, e_j(y)\right\rangle \nabla\phi_s(x-y) \cdot \omega_{ij}(y)\,dy.$$

We can estimate the second term with the same method as before. Furthermore, by Lemma 2.6, we have

$$[\phi_s(x-\cdot)(u - \bar{u}_{B_s(x)})]_{\mathrm{BMO}(B_s(x))} \le C_4 s^{1-m}\psi(x)$$

for a constant C_4 that depends only on m, N, and ζ (where ψ is the function defined on page 130). Thus we can apply Theorem 2.3 and Theorem 2.4 again. We obtain

$$\int_{B_s(x)} \phi_s(x-y)\left\langle u(y) - \bar{u}_{B_s(x)}, \nabla(\zeta(y)h_i^l(y)e_j(y))\right\rangle \cdot \omega_{ij}(y)\,dy$$

$$\le C_5 t^{1-m}\psi(x)\|\omega\|_{L^2(B_s(x))}\|\nabla e\|_{L^2(B_s(x))}$$

$$\le C_6\psi(x)Z_0(x)\epsilon_0$$

for constants C_5 and C_6 that depend only on m, N, and ζ. We know that (4.28) and (4.29) hold, thus the right-hand side is in $L^1(B_{1/4}(0))$ with

$$\|\psi Z_0\|_{L^1(B_{1/4}(0))} \le C\left(\sqrt{\epsilon} + \|f\|_{L^{m/2}(B_1(0))}\right)\left(\sqrt{\epsilon+\epsilon_0} + \|f\|_{L^{m/2}(B_1(0))}\right).$$

To conclude the proof, we can now argue like in the proofs of Theorem 2.4 and Lemma 4.7. □

We are finally in a position to prove the energy inequality mentioned at the beginning of this section.

Proposition 4.2 *There exists a constant C, depending only on m and N, such that for any $\delta > 0$ there exist a number $\epsilon_1 > 0$, depending on m, N, and δ, with the following property. Suppose $\epsilon \le \epsilon_1$ and $u \in H^1(\Omega; N)$ is a solution of (4.1) for $f \in L^p(\Omega; \mathbb{R}^n)$, where $p = 2$ if $m \le 4$ and $p = \frac{m}{2}$ else. If $B_r(x_0) \subset \Omega$ is a ball such that*

$$\sup_{x_1 \in B_{r/2}(x_0)} \sup_{0<s\le r/2} \left(s^{2-m}\int_{B_s(x_1)} |\nabla u|^2\,dx\right) \le \epsilon,$$

then

$$\left(\frac{r}{8}\right)^{2-m}\int_{B_{r/8}(x_0)}|\nabla u|^2\,dx\le\delta\left(\epsilon+r^{2p-m}\|f\|^2_{L^p(B_r(x_0))}\right)$$
$$+\frac{C}{\delta}\fint_{B_r(x_0)}|u-\bar{u}_{B_r(x_0)}|^2\,dx. \quad (4.30)$$

Proof. We can assume that $B_r(x_0)=B_1(0)$. Choose a cut-off function $\zeta\in C_0^\infty(B_{1/4}(0))$ with $0\le\zeta\le 1$ and $\zeta\equiv 1$ on $B_{1/8}(0)$. An integration by parts shows that

$$\begin{aligned}\int_{B_{1/4}(0)}\zeta|\left\langle\nabla u,e_i\right\rangle|^2\,dx=&-\int_{B_{1/4}(0)}\left\langle u-\bar{u}_{B_1(0)},e_i\right\rangle\nabla\zeta\cdot\left\langle\nabla u,e_i\right\rangle\,dx\\&-\sum_{j=1}^{k}\int_{B_{1/4}(0)}\zeta\left\langle u-\bar{u}_{B_1(0)},e_i\right\rangle\left\langle\nabla u,e_j\right\rangle\cdot\omega_{ij}\,dx\\&-\int_{B_{1/4}(0)}\zeta\left\langle u-\bar{u}_{B_1(0)},e_i\right\rangle\left\langle f,e_i\right\rangle\,dx\\&-\int_{B_{1/4}(0)}\zeta\left\langle u-\bar{u}_{B_1(0)},\nabla e_i\right\rangle\cdot\left\langle\nabla u,e_i\right\rangle\,dx.\end{aligned}$$

Clearly we can estimate

$$\int_{B_{1/4}(0)}\left\langle u-\bar{u}_{B_1(0)},e_i\right\rangle\nabla\zeta\cdot\left\langle\nabla u,e_i\right\rangle\,dx\le\delta\epsilon+\frac{C}{\delta}\int_{B_1(0)}|u-\bar{u}_{B_1(0)}|^2\,dx,$$

using Young's inequality. Similarly,

$$\begin{aligned}-\int_{B_{1/4}(0)}\zeta\left\langle u-\bar{u}_{B_1(0)},e_i\right\rangle\left\langle f,e_i\right\rangle\,dx\le\;&\delta\|f\|^2_{L^p(B_1(0))}\\&+\frac{C}{\delta}\int_{B_1(0)}|u-\bar{u}_{B_1(0)}|^2\,dx.\end{aligned}$$

As we have done several times before, we can apply Theorem 2.3 and Theorem 2.4 to obtain the inequality

$$-\int_{B_{1/4}(0)}\zeta\left\langle u-\bar{u}_{B_1(0)},e_i\right\rangle\left\langle\nabla u,e_j\right\rangle\cdot\omega_{ij}\,dx\le C_1\epsilon^{3/2}$$

for a constant C_1 that depends only on m, N, and the choice of ζ. Finally, by the same arguments and Lemma 4.9, we have

$$-\int_{B_{1/4}(0)} \zeta \left\langle u - \bar{u}_{B_1(0)}, \nabla e_i \right\rangle \cdot \langle \nabla u, e_i \rangle \, dx$$
$$\leq C_2 \sqrt{\epsilon} \left(\sqrt{\epsilon} + \|f\|_{L^{m/2}(B_1(0))} \right) \left(\sqrt{\epsilon + \epsilon_0} + \|f\|_{L^{m/2}(B_1(0))} \right)$$

for another constant C_2 that depends only on m, N, and ζ. Here ϵ_0 is a number that can be chosen arbitrarily small. We have

$$|\nabla u|^2 = \sum_{i=1}^{k} |\langle \nabla u, e_i \rangle|^2$$

almost everywhere, thus we only have to combine these inequalities to prove the proposition. □

4.6 Small Dimensions

In this section we study almost harmonic maps in dimensions $m \leq 4$. We assume that u is a solution of (4.1) for some $f \in L^2(\Omega; \mathbb{R}^n)$. The aim is to show that under reasonable conditions, we can find estimates for the second derivatives of u. Some of the methods we use here could also be applied in higher dimensions in the case $f \in L^p(\Omega; \mathbb{R}^n)$ with $p > \frac{m}{2}$. In order to keep the presentation simple, however, we consider only small dimensions. Our aim is not to prove the most general result we can find, but rather to show how the arguments work. Thus throughout this section, we assume $m \leq 4$.

First we observe that we can improve the integrability of the gradient of an almost harmonic map with small scaled energy. For this purpose we combine the energy inequality of the last section with the Gehring type lemma of Section 2.6.

Lemma 4.10 *Suppose $m \leq 4$. There exist $q_0 > 2$ and $C, \epsilon > 0$, depending only on N, such that the following holds: Suppose $f \in L^2(B_r(x_0); \mathbb{R}^n)$, and suppose $u \in H^1(B_r(x_0); N)$ is a stationary solution of (4.1). If*

$$r^{2-m} \int_{B_r(x_0)} |\nabla u|^2 \, dx + r^{4-m} \int_{B_r(x_0)} |f|^2 \, dx \leq \epsilon, \tag{4.31}$$

then

$$r^{q_0-m} \int_{B_{r/4}(x_0)} |\nabla u|^{q_0} \, dx \leq C \epsilon^{q_0/2}.$$

Proof. We can assume that $B_r(x_0) = B_1(0)$, otherwise we rescale everything. We note that (4.31) and the monotonicity formula in Lemma 4.2 imply

$$\sup_{x_1 \in B_{3/4}(0)} \sup_{0<r\leq 1/4} \left(r^{2-m} \int_{B_r(x_1)} |\nabla u|^2 \, dx \right) \leq C\epsilon.$$

Thus for any given $\delta > 0$, we can choose ϵ so small that Proposition 4.2 can be applied. Hence estimate (4.30) holds for any $x_0 \in B_{1/2}(0)$ and any $r \in (0, \frac{1}{4})$. Using a simply covering argument and applying the monotonicity inequality once more, we see that

$$r^{2-m} \int_{B_{r/2}(x_0)} |\nabla u|^2 \, dx \leq C\delta r^{2-m} \left(\int_{B_{2r}(x_0)} |\nabla u|^2 \, dx + r^2 \|f\|^2_{L^2(B_{2r}(x_0))} \right) + \frac{C}{\delta} \fint_{B_{2r}(x_0)} |u - \bar{u}_{B_r(x_0)}|^2 \, dx.$$

Set

$$q = \frac{2m}{m+2}.$$

Estimating the last term of the inequality above with the Sobolev inequality, we obtain

$$\fint_{B_{r/2}(x_0)} |\nabla u|^2 \, dx \leq C\delta \left(\fint_{B_{2r}(x_0)} |\nabla u|^2 \, dx + r^{2-m} \|f\|^2_{L^2(B_{2r}(x_0))} \right) + \frac{C}{\delta} \left(\fint_{B_{2r}(x_0)} |\nabla u|^q \, dx \right)^{\frac{2}{q}}.$$

Note that

$$r^{2-m} \|f\|^2_{L^2(B_{2r}(x_0))} \leq \left(\int_{B_r(x_0)} |f|^2 \, dx \right)^{2/m} \left(\omega_m \fint_{B_{2r}(x_0)} |f|^2 \, dx \right)^{(m-2)/m} \leq C\epsilon^{2/m} \fint_{B_{2r}(x_0)} F^2 \, dx$$

for the function

$$F(x) = \sup_{r>0} \left(\fint_{B_{4r}(x) \cap B_1(0)} |f|^2 \, dx \right)^{\frac{m-2}{2m}},$$

which satisfies

$$\|F\|_{L^3(B_1(0))} \le C\|f\|_{L^2(B_1(0))}^{(m-2)/m}$$

by the Hardy-Littlewood maximal theorem. Thus we have

$$\begin{aligned}\fint_{B_{r/2}(x_0)} |\nabla u|^2\,dx \le C\delta\left(\fint_{B_{2r}(x_0)} |\nabla u|^2\,dx + \epsilon^{2/m}\fint_{B_{2r}(x_0)} F^2\,dx\right)\\ + \frac{C}{\delta}\left(\fint_{B_{2r}(x_0)} |\nabla u|^q\,dx\right)^{2/q}\end{aligned}$$

for every ball $B_r(x_0) \subset B_{1/2}(0)$. If δ is chosen appropriately, we can apply (a rescaled version of) Lemma 2.11 to the function $f = |\nabla u|^q$ and find that the estimate

$$\int_{B_{1/4}(0)} |\nabla u|^{q_0}\,dx \le C\epsilon^{q_0/2}$$

holds for a number $q_0 > 2$. □

Lemma 4.11 *Suppose $m \le 4$. For any $q_0 > 2$ there exist two constants C_0 and $\epsilon > 0$, depending only on N and q_0, with the following property: Let $f \in L^2(B_1(0);\mathbb{R}^n)$. Suppose $q_0 \le q \le 4$, and suppose $u \in H^1(B_1(0);N) \cap W^{1,q}(B_1(0);N)$ is a stationary solution of (4.1) such that*

$$\int_{B_1(0)} (|\nabla u|^2 + |f|^2)\,dx \le \epsilon. \tag{4.32}$$

Then

$$\begin{aligned}\fint_{B_{r/2}(x_0)} |\nabla u|^q\,dx \le C_0\left(\fint_{B_r(x_0)} (|\nabla u|^2 + r^2|f|^2)\,dx\right)^{q/2}\\ + C_0\epsilon^{q/4}\fint_{B_r(x_0)} |\nabla u|^q\,dx\end{aligned}$$

for any ball $B_r(x_0) \subset B_{1/2}(0)$. In particular,

$$\int_{B_{r/2}(x_0)} |\nabla u|^q\,dx \le C_0\left(\epsilon^{q/2} r^{m-q} + \epsilon^{q/4}\int_{B_r(x_0)} |\nabla u|^q\,dx\right)$$

for any such ball.

Proof. Choose $\eta \in C_0^\infty(B_{3/4}(0))$ with $\eta \equiv 1$ in $B_{1/2}(0)$. For $x_0 \in \mathbb{R}^m$ and $r > 0$, set

$$\eta_{x_0,r}(x) = \eta\left(\frac{x - x_0}{r}\right).$$

If $B_r(x_0) \subset B_{1/2}(0)$, consider the map

$$v = \eta_{x_0,r}(u - \bar{u}_{B_r(x_0)}).$$

According to Lemma 2.6 we have

$$[v]_{\mathrm{BMO}(\mathbb{R}^m)} \leq C_1[u]_{\mathrm{BMO}(B_{3r/4}(0))}$$

for a constant C_1 that depends only on n and the choice of η. Because of Lemma 4.2 and the Poincaré inequality, this implies

$$[v]_{\mathrm{BMO}(\mathbb{R}^m)} \leq C_2\left(r^{1-m/2}\|\nabla u\|_{L^2(B_r(x_0))} + r^{2-m/2}\|f\|_{L^2(B_r(x_0))}\right), \quad (4.33)$$

where C_2 depends on n and the choice of η. Moreover, we have

$$\|\Delta v\|_{L^{q/2}(\mathbb{R}^m)} \leq C_3\left(r^{-1}\|\nabla u\|_{L^{q/2}(B_r(x_0))} + \|\Delta u\|_{L^{q/2}(B_r(x_0))}\right) \leq C_4 I,$$

for constants C_3 and C_4 that depend only on N and η, where

$$I = r^{-1}\|\nabla u\|_{L^{q/2}(B_r(x_0))} + \|f\|_{L^{q/2}(B_r(x_0))} + \|\nabla u\|^2_{L^q(B_r(x_0))}.$$

According to standard L^p-estimates for elliptic equations, we thus have

$$\|\nabla^2 v\|_{L^{q/2}(\mathbb{R}^m)} \leq C_5 I$$

for another constant C_5 that depends only on N and η. Combining this inequality with (4.33) and the Gagliardo-Nirenberg type inequality of Proposition 2.2, we obtain

$$\|\nabla v\|^2_{L^q(\mathbb{R}^m)} \leq C_6 I\left(r^{1-m/2}\|\nabla u\|_{L^2(B_r(x_0))} + r^{2-m/2}\|f\|_{L^2(B_r(x_0))}\right)$$

for a constant C_6 that depends only on N and η. Since $q/2 \leq 2$, it follows that

$$\begin{aligned}
&\|\nabla u\|^q_{L^q(B_{r/2}(x_0))} \\
&\quad \leq C_7 r^{-q}\|u - \bar{u}_{B_r(x_0)}\|^q_{L^q(B_r(x_0))} + C_7 r^{m-qm/2}\|\nabla u\|^q_{L^2(B_r(x_0))} \\
&\qquad + C_7 r^{m+q-qm/2}\|f\|^q_{L^2(B_r(x_0))} \\
&\qquad + C_7 r^{q/2-mq/4}\left(\|\nabla u\|_{L^2(B_r(x_0))} + r\|f\|_{L^2(B_r(x_0))}\right)^{q/2}\|\nabla u\|^q_{L^q(B_r(x_0))},
\end{aligned}$$

where C_7 depends on N and η. We use the Sobolev inequality to estimate

$$\|u - \bar{u}_{B_r(x_0)}\|^q_{L^q(B_r(x_0))} \leq C_8 r^{m-qm/2} \|\nabla u\|^q_{L^2(B_r(x_0))}$$

for another constant C_8 that depends only on n. Lemma 4.2 implies furthermore, that

$$r^{1-m/2}\|\nabla u\|_{L^2(B_r(x_0))} + r^{2-m/2}\|f\|_{L^2(B_r(x_0))} \leq C\sqrt{\epsilon}. \tag{4.34}$$

Combining these estimates, we obtain the first inequality of the lemma, if ϵ is small enough. Using (4.34) once more, we see that the second inequality follows from the first. □

Lemma 4.12 *Suppose $m \leq 4$. For any $q_0 > 2$ there exist two constants C_0 and $\epsilon_0 > 0$, depending only on N and q_0, with the following property: Let $f \in L^2(B_1(0);\mathbb{R}^n)$. Suppose $q_0 \leq q \leq 4$, and suppose $u \in H^1(B_1(0);N) \cap W^{1,q}(B_1(0);N)$ is a stationary solution of (4.1) satisfying*

$$\int_{B_1(0)} (|\nabla u|^2 + |f|^2)\, dx = \epsilon \leq \epsilon_0. \tag{4.35}$$

Then

$$\int_{B_{1/4}(0)} |\nabla u|^q \, dx \leq C_0 \epsilon^{q/2}.$$

Proof. There exists a countable collection of balls $\{B_{r_i}(x_i)\}_{i\in\mathbb{N}}$ in $B_{1/2}(0)$, satisfying

$$2r_i \leq \frac{1}{2} - |x_i| \leq 8r_i$$

for each i, such that

$$B_{1/2}(0) = \bigcup_{i=1}^{\infty} B_{r_i/2}(x_i),$$

but each point of $B_{1/2}(0)$ is contained in at most k_0 of the balls $B_{r_i}(x_i)$ for a universal constant k_0. Define

$$g(x) = \left(\frac{1}{2} - |x|\right)^q |\nabla u|^q, \quad x \in B_{1/2}(0),$$

then the second inequality of Lemma 4.11 implies that

$$\int_{B_{r_i/2}(x_i)} g\, dx \leq C_1 \left(\epsilon^{q/2}|B_{r_i}(x_i)| + \epsilon^{q/4} \int_{B_{r_i}(x_i)} g\, dx\right)$$

for each $i \in \mathbb{N}$, where C_1 is a constant that depends only on q_0 and N. Summing over i, we obtain

$$\int_{B_{1/2}(0)} g\,dx \le C_1 k_0 \left(\epsilon^{q/2} + \epsilon^{q/4} \int_{B_{1/2}(0)} g\,dx \right).$$

If ϵ_0 is sufficiently small, the claim follows. □

Combining Lemmas 4.10–4.12, we are now able to prove the following proposition.

Proposition 4.3 *Suppose $m \le 4$. There exist two numbers C and $\epsilon > 0$, depending only on N, that have the following property. Suppose $f \in L^2(B_r(x_0); \mathbb{R}^n)$, and suppose $u \in H^1(B_r(x_0); N)$ is a stationary solution of (4.1) such that*

$$r^{2-m} \int_{B_r(x_0)} (|\nabla u|^2 + r^2|f|^2)\,dx \le \epsilon. \tag{4.36}$$

Then $u \in H^2(B_{r/2}(x_0); N) \cap W^{1,4}(B_{r/2}(x_0); N)$ with

$$r^{4-m} \int_{B_{r/2}(x_0)} |\nabla^2 u|^2\,dx + \left(r^{4-m} \int_{B_{r/2}(x_0)} |\nabla u|^4\,dx \right)^{1/2} \le C r^{2-m} \int_{B_r(x_0)} (|\nabla u|^2 + r^2|f|^2)\,dx.$$

Proof. We can assume that $B_r(x_0) = B_1(0)$. Moreover, it suffices to show that

$$\int_{B_{1/8}(0)} |\nabla^2 u|^2\,dx + \left(\int_{B_{1/8}(0)} |\nabla u|^4\,dx \right)^{1/2} \le C \int_{B_1(0)} (|\nabla u|^2 + r^2|f|^2)\,dx.$$

The claim of the proposition then follows by a simple covering argument.

To prove this inequality, we proceed as follows. According to Lemma 4.2, we have

$$r^{2-m} \int_{B_r(x_0)} |\nabla u|^2\,dx \le C\epsilon$$

for any $x_0 \in B_{1/2}(0)$ and any $r \in (0, \frac{1}{2}]$. In particular the inequalities (4.32) and (4.35) in Lemma 4.11 and Lemma 4.12 are satisfied for the rescaled maps $u(rx + x_0)$, provided that ϵ is sufficiently small. We will use this fact later when we apply these lemmas.

First, however, we use Lemma 4.10 to show that

$$\nabla u \in L^{q_0}(B_{1/2}(0); \mathbb{R}^{m\times n})$$

for some $q_0 > 2$. Define

$$q_1 = \sup\left\{q \in [2,4] : \ \nabla u \in L^q_{\mathrm{loc}}(B_{1/2}(0); \mathbb{R}^{m\times n})\right\}.$$

Then $q_1 \geq q_0$. If $q_1 > q_0$, then Lemma 4.12 gives bounds for

$$\int_{B_{r/4}(x_0)} |\nabla u|^q \, dx$$

that are uniform in $q \in [q_0, q_1)$, whenever $B_r(x_0) \subset B_{1/2}(0)$. It follows that

$$\nabla u \in L^{q_1}_{\mathrm{loc}}(B_{1/2}(0); \mathbb{R}^{m\times n}).$$

Assume now that $q_1 < 4$. Then we can apply Lemma 4.11 for $q = q_1$ (provided that ϵ is sufficiently small). We observe that

$$r^2 \fint_{B_r(x_0)} |f|^2 \, dx \leq C\sqrt{\epsilon} \left(\fint_{B_r(x_0)} |f|^2 \, dx \right)^{1/2} \leq C\sqrt{\epsilon} \fint_{B_r(x_0)} G \, dx$$

for the function

$$G(x) = \sup_{r>0} \left(\fint_{B_{1/2}(0)\cap B_{2r}(x)} |f|^2 \, dx \right)^{1/2},$$

which is in $L^s(\mathbb{R}^m)$ for every $s < 2$ by the Hardy-Littlewood maximal theorem. Thus by Lemma 4.11, we have

$$\fint_{B_{r/2}(x_0)} |\nabla u|^{q_1} \, dx \leq C \left(\fint_{B_r(x_0)} |\nabla u|^2 \, dx \right)^{q_1/2} + C\epsilon^{q_1/4} \fint_{B_r(x_0)} |\nabla u|^{q_1} \, dx + C\epsilon^{q_1/4} \fint_{B_r(x_0)} G^{q_1/2} \, dx.$$

According to Lemma 2.11, this inequality implies

$$\nabla u \in L^{q_2}_{\mathrm{loc}}(B_{1/2}(0); \mathbb{R}^{m\times n})$$

for some $q_2 > q_1$, provided that ϵ is small enough, in contradiction to the definition of q_1. Hence $q_1 = 4$. It follows in particular that

$$\nabla u \in L^4_{\mathrm{loc}}(B_{1/2}(0); \mathbb{R}^{m\times n}).$$

Applying Lemma 4.11 again, we infer

$$\left(\int_{B_{1/4}(0)} |\nabla u|^4 \, dx \right)^{1/2} \leq C r^{2-m} \int_{B_r(x_0)} (|\nabla u|^2 + r^2 |f|^2) \, dx.$$

Standard elliptic estimates then imply

$$\int_{B_{1/8}(0)} |\nabla^2 u|^2 \, dx \leq C r^{2-m} \int_{B_r(x_0)} (|\nabla u|^2 + r^2 |f|^2) \, dx.$$

This concludes the proof. □

Finally, if the condition (4.36) is not satisfied, we can still find a small set such that we have good estimates away from this set. To this end, we use the same covering arguments as in the proof of the previous results of this kind (that is, arguments as described in the proof of Theorem 3.2). We leave the details to the reader.

Theorem 4.2 *Suppose* $m \leq 2$ *and* $f \in L^2(\Omega; \mathbb{R}^n)$. *If* $u \in H^1(\Omega; N)$ *is a stationary solution of (4.1), then there exists a relatively closed set* $\Sigma \subset \Omega$ *of vanishing* $(m-2)$*-dimensional Hausdorff measure such that* $u \in H^2_{\text{loc}}(\Omega \backslash \Sigma; N) \cap W^{1,4}_{\text{loc}}(\Omega \backslash \Sigma; N)$.

Chapter 5

Evolution Problems

It was first observed by M. Feldman [1994], and independently shown by Y. Chen, J. Li, and F.-H. Lin [1995], that the methods from the regularity theory of harmonic maps can be used to prove partial regularity for solutions of a parabolic problem, namely the harmonic map heat flow, at least if the target manifold is a sphere. We will discuss two parabolic problems in this chapter for which the methods of the previous chapters give regularity results. One of them is the harmonic map heat flow, the other is the Landau-Lifshitz equation from the theory of ferromagnetism.

In order to obtain partial regularity results for such problems, we first need to find an appropriate generalization of the condition defining stationary weakly harmonic maps. For the harmonic map heat flow, a stability condition suggested by M. Feldman [1994] can be used. This condition implies in particular a parabolic version of the monotonicity formula that played an important role in the previous chapters. If the target manifold is $\mathbb{S}^{n-1}$, we can then prove parabolic versions of the results of Section 3.5. For general targets, the situation is more difficult. There is no obvious way to apply the methods of Section 3.7 to parabolic problems. But if the dimension of the domain Ω is small (more precisely, $m \leq 4$), we can use the results of Chapter 4 instead, and then we can find regularity results nevertheless. For $m \geq 5$, the problem is open.

The difficulties for the Landau-Lifshitz equation are of a different nature. Here we lack a suitable parabolic monotonicity formula. If $m \leq 4$, however, we can replace it by the formula from the elliptic theory. Thus although the problem is different, the remedy is the same as for the harmonic map heat flow: We consider only domains of small dimensions.

5.1 The Stability Condition for the Harmonic Map Heat Flow

We first study the harmonic map heat flow

$$\frac{\partial u}{\partial t} = \Delta u + A(u)(\nabla u, \nabla u) \quad \text{in } (0,T) \times \Omega. \tag{5.1}$$

Here and throughout the chapter T is a fixed positive number or infinity. We observe that any weakly harmonic map $u_0 \in H^1(\Omega; N)$ provides a weak solution of (5.1) of the form $u(t,x) = u_0(x)$. In order to prove a regularity result for solutions of the harmonic map heat flow, we thus need additional assumptions again.

For the elliptic problems, the monotonicity formula was the key to regularity. Can we expect a similar formula for the harmonic map heat flow? What would it look like? To answer the second question, we first assume that $\Omega = \mathbb{R}^m$ and that we have a smooth solution

$$u \in C^\infty([-1,0) \times \mathbb{R}^m; N) \cap L^\infty([-1,0); H^1(\mathbb{R}^m; N))$$

of (5.1). It was observed by M. Struwe [1988] that in this situation, it is a good idea to study the quantity

$$\Phi(u;r) = \frac{r^2}{2} \int_{\mathbb{R}^m \times \{-r^2\}} |\nabla u|^2 G \, dx, \quad 0 < r \leq 1,$$

where G is the fundamental solutions of the backward heat equation in $\mathbb{R}^m$, that is,

$$G(t,x) = \frac{1}{(4\pi|t|)^{m/2}} \exp\left(-\frac{|x|^2}{4|t|}\right), \quad t < 0.$$

Writing $u_r(t,x) = u(r^2 t, rx)$, we can easily verify that

$$\Phi(u;r) = \Phi(u_r;1).$$

Thus

$$\begin{aligned}
\frac{d}{dr}\Phi(u;r) = \frac{d}{dr}\Phi(u_r;1) &= \int_{\mathbb{R}^m \times \{-1\}} \nabla u \cdot \nabla \left(\frac{\partial u_r}{\partial r}\right) G \, dx \\
&= -\int_{\mathbb{R}^m \times \{-1\}} \left\langle \Delta u - \frac{x}{2} \cdot \nabla u, \frac{\partial u_r}{\partial r} \right\rangle G \, dx \\
&= 2 \int_{\mathbb{R}^m \times \{-1\}} \left| \frac{\partial u}{\partial t} - \frac{x}{2} \cdot \nabla u \right|^2 G \, dx \geq 0.
\end{aligned}$$

If $\Phi(u;1) \leq \epsilon$ for a certain number $\epsilon > 0$, we conclude that $\Phi(u;r) \leq \epsilon$ for any $r \in (0,1]$, and hence

$$r^{2-m} \int_{B_r(0)} |\nabla u(-r^2, x)|^2 \, dx \leq (4\pi)^{m/2} e^{1/4} \epsilon.$$

If the same inequality holds for all translations of u, we have in particular a good estimate for $[u(t, \cdot)]_{\mathrm{BMO}(\mathbb{R}^m)}$ for all $t \in (-1, 0)$. Such an estimate would certainly be useful if we want to apply the methods of the previous chapters.

Of course, if we want to prove regularity, we cannot assume that we have smooth solutions. Therefore we need a reasonable weaker condition which implies this or a similar formula. Because we want to study other domains than $\mathbb{R}^m$, we also need a localized version of the formula.

We now define the stability condition of M. Feldman [1994]. We consider variations of the domain $(0,T) \times \Omega$ of the following kind: Suppose $\psi \in C_0^\infty((0,T) \times \Omega; \mathbb{R}^m)$ and $\tau \in C_0^\infty((0,T) \times \Omega; [0,\infty))$. If $u \in H^1((0,T) \times \Omega; N)$, we consider the family of maps

$$u_s(t,x) = u(t + s\tau(t,x), x + s\psi(t,x)). \tag{5.2}$$

Definition 5.1 A solution $u \in H^1((0,T) \times \Omega; N)$ of (5.1) is said to satisfy the *stability condition* if

$$\int_0^T \int_\Omega \left\langle \frac{\partial u}{\partial t}, \frac{\partial u_s}{\partial s}\bigg|_{s=0} \right\rangle dx\, dt + \frac{d}{ds}\bigg|_{s=0} \int_0^T E(u_s)\, dt \leq 0 \tag{5.3}$$

for all variations of u of the form (5.2).

It is not clear if this is the optimal condition to consider. In fact it is unknown in general if there exist solutions of the harmonic map heat flow satisfying the stability condition for any given initial data. The analogy to the elliptic problem, however, suggests that this is a reasonable condition. In any case, it is a condition we can work with and which provides regularity results. Once we know what properties we need, we can also replace it by something we might find more suitable for a given situation. For instance, instead of considering solutions that satisfy the stability condition, we can require that a solution satisfies the local monotonicity inequality which we will derive in this section.

We now study the consequences of the stability condition. For $\psi \in C_0^\infty((0,T) \times \Omega; \mathbb{R}^m)$ and $\tau \in C_0^\infty((0,T) \times \Omega; [0,\infty))$ as above, and for s in a small interval around 0, define the maps

$$\Phi_s(t,x) = (t + s\tau(t,x), x + s\psi(t,x))$$

on $(0,T) \times \Omega$, so that $u_s = u \circ \Phi_s$. If $|s|$ is sufficiently small, this is a diffeomorphism of $(0,T) \times \Omega$ onto itself, and we can define the inverse $\Psi_s = \Phi_s^{-1}$. We have

$$\frac{d}{ds}\bigg|_{s=0} \det d\Phi_s = \frac{\partial \tau}{\partial t} + \operatorname{div} \psi,$$

where d denotes the derivative in time-space. Hence

$$\frac{d}{ds}\bigg|_{s=0} \det d\Psi_s = -\left(\frac{\partial \tau}{\partial t} + \operatorname{div} \psi\right).$$

Moreover, we have

$$\nabla u_s = \nabla u \circ \Phi_s + s\nabla\tau \frac{\partial u}{\partial t} \circ \Phi_s + s \sum_{\alpha=1}^{m} \nabla\psi^\alpha \frac{\partial u}{\partial x^\alpha} \circ \Phi_s$$

and

$$\begin{aligned}
\frac{1}{2}|\nabla u_s|^2 = \frac{1}{2}|\nabla u \circ \Phi_s|^2 &+ s\nabla\tau \cdot \left\langle \frac{\partial u}{\partial t} \circ \Phi_s, \nabla u \circ \Phi_s \right\rangle \\
&+ s \sum_{\alpha=1}^{m} \nabla\psi^\alpha \cdot \left\langle \frac{\partial u}{\partial x^\alpha} \circ \Phi_s, \nabla u \circ \Phi_s \right\rangle \\
&+ \frac{s^2}{2} \left| \nabla\tau \frac{\partial u}{\partial t} \circ \Phi_s + \sum_{\alpha=1}^{m} \nabla\psi^\alpha \frac{\partial u}{\partial x^\alpha} \circ \Phi_s \right|^2 .
\end{aligned}$$

Hence

$$\begin{aligned}
\frac{d}{ds}\bigg|_{s=0} \int_0^T E(u_s)\, dt = \frac{1}{2} \frac{d}{ds}\bigg|_{s=0} &\int_0^T \int_\Omega |\nabla u \circ \Phi_s|^2 \, dx\, dt \\
&+ \int_0^T \int_\Omega \nabla\tau \cdot \left\langle \frac{\partial u}{\partial t}, \nabla u \right\rangle dx\, dt \\
&+ \sum_{\alpha=1}^{m} \int_0^T \int_\Omega \nabla\psi^\alpha \cdot \left\langle \frac{\partial u}{\partial x^\alpha}, \nabla u \right\rangle dx\, dt,
\end{aligned}$$

and for the first term on the right-hand side, we find, using the substitution formula,

$$\frac{1}{2}\left.\frac{d}{ds}\right|_{s=0}\int_0^T\int_\Omega |\nabla u\circ\Phi_s|^2\,dx\,dt = \frac{1}{2}\left.\frac{d}{ds}\right|_{s=0}\int_0^T\int_\Omega |\nabla u|^2|\det d\Psi_s|\,dx\,dt$$
$$= -\frac{1}{2}\int_0^T\int_\Omega |\nabla u|^2\left(\frac{\partial\tau}{\partial t}+\operatorname{div}\psi\right)dx\,dt.$$

We conclude that

$$\left.\frac{d}{dt}\right|_{s=0}\int_0^T E(u_s)\,dt = -\frac{1}{2}\int_0^T\int_\Omega |\nabla u|^2\left(\frac{\partial\tau}{\partial t}+\operatorname{div}\psi\right)dx\,dt$$
$$+\int_0^T\int_\Omega \nabla\tau\cdot\left\langle\frac{\partial u}{\partial t},\nabla u\right\rangle dx\,dt$$
$$+\sum_{\alpha=1}^m\int_0^T\int_\Omega \nabla\psi^\alpha\cdot\left\langle\frac{\partial u}{\partial x^\alpha},\nabla u\right\rangle dx\,dt.$$

It is readily verified that

$$\int_0^T\int_\Omega\left\langle\frac{\partial u}{\partial t},\left.\frac{\partial u_s}{\partial s}\right|_{s=0}\right\rangle dx\,dt = \int_0^T\int_\Omega\left(\tau\left|\frac{\partial u}{\partial t}\right|^2+\psi\cdot\left\langle\frac{\partial u}{\partial t},\nabla u\right\rangle\right)dx\,dt.$$

Hence we have derived the following inequality.

Lemma 5.1 *A map $u\in H^1((0,T)\times\Omega;N)$ satisfies the condition (5.3) if and only if*

$$\int_0^T\int_\Omega\left(\tau\left|\frac{\partial u}{\partial t}\right|^2+\psi\cdot\left\langle\frac{\partial u}{\partial t},\nabla u\right\rangle-\frac{1}{2}|\nabla u|^2\left(\frac{\partial\tau}{\partial t}+\operatorname{div}\psi\right)\right.$$
$$\left.+\nabla\tau\cdot\left\langle\frac{\partial u}{\partial t},\nabla u\right\rangle+\sum_{\alpha=1}^m\nabla\psi^\alpha\cdot\left\langle\frac{\partial u}{\partial x^\alpha},\nabla u\right\rangle\right)dx\,dt\le 0. \tag{5.4}$$

Remark 5.1 *If we choose $\psi=0$ and $\tau(t,x)=\eta_k(t)\xi(x)$ in (5.4) for some $\xi\in C_0^\infty(\Omega;[0,\infty))$ and a sequence of smooth functions η_k approximating the characteristic function of an interval $[t_1,t_2]\subset(0,T)$, we obtain in the limit*

$$\frac{1}{2}\int_\Omega\xi(x)(|\nabla u(t_2,x)|^2-|\nabla u(t_1,x)|^2)\,dx+\int_{t_1}^{t_2}\int_\Omega\xi\left|\frac{\partial u}{\partial t}\right|^2 dx\,dt$$
$$\le-\int_{t_1}^{t_2}\int_\Omega\nabla\xi\cdot\left\langle\frac{\partial u}{\partial t},\nabla u\right\rangle dx\,dt. \tag{5.5}$$

This is a localized version of the energy inequality (1.9) from the introduction. Such local energy inequalities are typical for the solutions of gradient flows. Unlike the corresponding inequalities for linear parabolic equations or certain other gradient flows, however, the local energy inequality for the harmonic map heat flow does not follow from the weak version of equation (5.1). We need the stability condition to derive it.

Lemma 5.2 *Let $u \in H^1((0,T)\times\Omega; N)$ be a map satisfying (5.3) for all admissible functions τ and ψ. Suppose $z_0 = (t_0, x_0) \in (0,T)\times\Omega$ and $r>0$ are such that $P_r(z_0) \subset (0,T)\times\Omega$. Then for every $t \in (t_0 - r^2/4, t_0 + r^2)$,*

$$r^{2-m}\int_{B_{r/2}(x_0)} |\nabla u(t,x)|^2\,dx + r^{2-m}\int_{P_{r/2}(z_0)} \left|\frac{\partial u}{\partial t}\right|^2 dz \leq Cr^{-m}\int_{P_r(z_0)} |\nabla u|^2\,dz$$

for a constant C that depends only on m.

Proof. In inequality (5.5), we choose $\xi = \zeta^2$, where $\zeta \in C_0^\infty(B_r(x_0))$ is a cut-off function satisfying $\zeta \equiv 1$ in $B_{r/2}(x_0)$ and $|\nabla\zeta| \leq 4/r$. Then for any pair of numbers t_1, t_2 with $t_0 - r^2 < t_1 < t_2 < t_0 + r^2$ we have

$$\begin{aligned}\frac{1}{2}\int_\Omega \zeta^2(x)(|\nabla u(t_2,x)|^2 - |\nabla u(t_1,x)|^2)\,dx &+ \int_{t_1}^{t_2}\int_\Omega \zeta^2\left|\frac{\partial u}{\partial t}\right|^2 dx\,dt \\ &\leq -2\int_{t_1}^{t_2}\int_\Omega \zeta\nabla\zeta\cdot\left\langle\frac{\partial u}{\partial t}, \nabla u\right\rangle dx\,dt \\ &\leq \frac{1}{2}\int_{t_1}^{t_2}\int_\Omega \zeta^2\left|\frac{\partial u}{\partial t}\right|^2 dx\,dt + 2\int_{P_r(z_0)} |\nabla\zeta|^2|\nabla u|^2\,dz.\end{aligned}$$

There exists a number $t_1 \in (t_0 - r^2, t_0 - r^2/4)$ such that

$$r^{2-m}\int_\Omega \zeta^2|\nabla u(t_1,x)|^2\,dx \leq Cr^{-m}\int_{P_r(z_0)} |\nabla u|^2\,dz,$$

and the claim follows immediately. □

We also want to derive a monotonicity inequality, of the type mentioned at the beginning of this section, for maps satisfying the stability condition. To this end, we consider the fundamental solution of the backward heat equation with singularity at the point $z_0 = (t_0, x_0)$,

$$G_{z_0}(t,x) = \frac{1}{(4\pi(t_0-t))^{m/2}}\exp\left(-\frac{|x-x_0|^2}{4(t_0-t)}\right), \quad t<t_0,\ x\in\mathbb{R}^m.$$

To find a localized monotonicity inequality, we multiply G_{z_0} with a cut-off function and insert a τ and a ψ derived from this product as test functions into (5.4). We eventually find the following inequality.

Lemma 5.3 *Suppose $u \in H^1((0,T)\times\Omega;N)$ satisfies (5.3) for all admissible functions τ and ψ. Let $z_0 = (t_0,x_0) \in (0,T)\times\Omega$ and $r > 0$ such that $B_r(x_0) \subset \Omega$ and let $0 < t_1 < t_2 < t_0$ such that $t_1 \geq t_0 - r^2$. Suppose $\xi \in C_0^\infty(B_r(x_0);[0,\infty))$ is a function satisfying $\xi \equiv 1$ in $B_{r/2}(x_0)$ and $|\nabla\xi|^2 + |\nabla^2\xi| \leq cr^{-2}$. Then*

$$
\begin{aligned}
(t_0-t_2)\int_{\Omega\times\{t_2\}} \xi|\nabla u|^2 G_{z_0}\,dx - (t_0-t_1)\int_{\Omega\times\{t_1\}} \xi|\nabla u|^2 G_{z_0}\,dx& \\
+\int_{t_1}^{t_2}(t_0-t)\int_\Omega \left|2\frac{\partial u}{\partial t} + \frac{x-x_0}{t-t_0}\cdot\nabla u\right|^2 \xi G_{z_0}\,dx\,dt& \\
\leq C_0 r^{-m}\int_{t_1}^{t_2}\int_{B_r(x_0)} \left(|\nabla u|^2 + r^2\left|\frac{\partial u}{\partial t}\right|^2\right)dx\,dt& \qquad (5.6)
\end{aligned}
$$

for a constant C_0 that depends only on c.

Proof. Let $\zeta \in C_0^\infty((0,t_0);[0,\infty))$. We insert the functions

$$\tau(t,x) = (t_0-t)\zeta(t)\xi(x)G_{z_0}(t,x)$$

and

$$\psi(t,x) = \nabla\tau(t,x) = \zeta(t)\left((t_0-t)\nabla\xi(x) - \frac{1}{2}(x-x_0)\xi(x)\right)G_{z_0}(t,x)$$

into (5.4). We have

$$
\begin{aligned}
\frac{\partial\tau}{\partial t}(t,x) + \operatorname{div}\psi(t,x) = \big[((t_0-t)\zeta'(t) - \zeta(t))\xi(x) + \zeta(t)(t_0-t)\Delta\xi(x)& \\
- \zeta(t)(x-x_0)\cdot\nabla\xi(x))\big]G_{z_0}(t,x)&
\end{aligned}
$$

and

$$
\begin{aligned}
\frac{\partial\psi^\alpha}{\partial x^\beta}(t,x) = \zeta(t)\Bigg((t_0-t)\frac{\partial^2\xi(x)}{\partial x^\alpha\partial x^\beta} - \frac{1}{2}\delta_{\alpha\beta}\xi(x) - \frac{1}{2}(x^\alpha - x_0^\alpha)\frac{\partial\xi(x)}{\partial x^\beta}& \\
- \frac{1}{2}(x^\beta - x_0^\beta)\frac{\partial\xi(x)}{\partial x^\alpha} + \frac{(x^\alpha-x_0^\alpha)(x^\beta-x_0^\beta)}{4(t_0-t)}\xi(x)\Bigg)G_{z_0}(t,x).&
\end{aligned}
$$

Hence by (5.4),

$$\int_0^T \int_\Omega \Bigg((t_0 - t)\zeta\xi \left|\frac{\partial u}{\partial t}\right|^2 - \frac{1}{2}(t_0 - t)\zeta'\xi|\nabla u|^2 - \zeta\xi \left\langle \frac{\partial u}{\partial t}, (x - x_0)\cdot \nabla u \right\rangle + \frac{\zeta\xi}{4(t_0 - t)}|(x - x_0)\cdot \nabla u|^2 \Bigg) G_{z_0}\, dx\, dt$$

$$\le \int_0^T \int_\Omega \Bigg(\frac{1}{2}(t - t_0)\zeta\Delta\xi|\nabla u|^2 - 2(t_0 - t)\zeta\nabla\xi \cdot \left\langle \frac{\partial u}{\partial t}, \nabla u \right\rangle - \frac{1}{2}\zeta(x - x_0)\cdot\nabla\xi|\nabla u|^2 - (t_0 - t)\zeta \sum_{\alpha,\beta=1}^m \frac{\partial^2 \xi}{\partial x^\alpha \partial x^\beta} \left\langle \frac{\partial u}{\partial x^\alpha}, \frac{\partial u}{\partial x^\beta} \right\rangle + \zeta \left\langle \nabla\xi\cdot\nabla u, (x - x_0)\cdot\nabla u \right\rangle \Bigg) G_{z_0}\, dx\, dt.$$

Letting ζ approximate the characteristic function of the interval $[t_1, t_2]$, we see that

$$\begin{aligned}
&(t_0 - t_2)\int_{\Omega\times\{t_2\}} \xi|\nabla u|^2 G_{z_0}\, dx - (t_0 - t_1)\int_{\Omega\times\{t_1\}} \xi|\nabla u|^2 G_{z_0}\, dx \\
&+ 2\int_{t_1}^{t_2}\int_\Omega \Bigg((t_0 - t)\left|\frac{\partial u}{\partial t}\right|^2 - \left\langle \frac{\partial u}{\partial t}, (x - x_0)\cdot\nabla u \right\rangle + \frac{|(x - x_0)\cdot\nabla u|^2}{4(t_0 - t)} \Bigg) \xi G_{z_0}\, dx\, dt \\
&\le C_1 \int_{t_1}^{t_2} \int_{B_r(x_0)\setminus B_{r/2}(x_0)} \Bigg[(t_0 - t)\left(\left|\frac{\partial u}{\partial t}\right|^2 + (|\nabla\xi|^2 + |\nabla^2\xi|)|\nabla u|^2 \right) + |x - x_0||\nabla\xi||\nabla u|^2 \Bigg] G_{z_0}\, dx\, dt
\end{aligned} \tag{5.7}$$

for a constant C_1 that depends only on c. On $(B_r(x_0)\setminus B_{r/2}(x_0)) \times (t_0 - r^2, t_0)$, we have

$$G_{z_0} \le Cr^{-m}, \quad |t_0 - t|G_{z_0} \le C_2 r^{2-m}, \quad \text{and} \quad |x - x_0|G_{z_0} \le C_2 r^{1-m}.$$

Hence the right-hand side of (5.7) is bounded by

$$C_2 r^{-m} \int_{t_1}^{t_2} \int_{B_r(x_0)} \left(|\nabla u|^2 + r^2 \left| \frac{\partial u}{\partial t} \right|^2 \right) dx\, dt,$$

where C_2 is a constant that depends only on c. Observing that

$$(t_0 - t) \left| \frac{\partial u}{\partial t} \right|^2 - \left\langle \frac{\partial u}{\partial t}, (x - x_0) \cdot \nabla u \right\rangle + \frac{|(x - x_0) \cdot \nabla u|^2}{4(t_0 - t)} = (t_0 - t) \left| \frac{\partial u}{\partial t} + \frac{x - x_0}{2(t - t_0)} \cdot \nabla u \right|^2,$$

we conclude that (5.6) holds. □

Lemma 5.4 *Let $u \in H^1((0,T) \times \Omega; N)$ be a map that satisfies (5.3) for all admissible functions τ and ψ. If $P_r(z_0) \subset (0,T) \times \Omega$, then*

$$s^{2-m} \left(\int_{B_s(x_1)} |\nabla u(t_1, x)|^2\, dx + \int_{P_s(z_1)} \left| \frac{\partial u}{\partial t} \right|^2 dz \right) \le C r^{-m} \int_{P_r(z_0)} |\nabla u|^2\, dz$$

for all $z_1 = (t_1, x_1) \in P_{r/2}(z_0)$ and all $s \in (0, r/2]$, where C is a universal constant.

Proof. This inequality now follows easily from Lemma 5.2 and Lemma 5.3. We only have to take into account that

$$(t_0 - t) G_{z_0} \ge C r^{2-m}$$

on the set

$$\left\{ z = (t, x) \in \mathbb{R} \times \mathbb{R}^m : \ t_0 - r^2 < t < t_0 - r^2/4, \ |x - x_0| < r \right\}.$$

Then we use an appropriate cut-off functions and similar arguments as in the proof of Lemma 5.2. □

5.2 Higher Regularity for Evolution Problems

When we want to prove regularity of solutions of either the harmonic map heat flow or the Landau-Lifshitz equation, the greatest challenge again is to prove that the solutions are continuous. Similarly as for the elliptic problem, higher regularity then follows relatively easily. We show in this chapter how the arguments of Section 3.4 can be modified so that they prove higher regularity for continuous solutions of the evolution equations.

We treat the harmonic map heat flow and the Landau-Lifshitz equation simultaneously in most of this section.

The fact that the Landau-Lifshitz equation

$$a\,\frac{\partial u}{\partial t} + b\,u \wedge \frac{\partial u}{\partial t} = \Delta u + |\nabla u|^2 u \quad \text{in } (0,T)\times\Omega$$

has nonlinearities in a term of leading order (namely the expression $b\,u\wedge\frac{\partial u}{\partial t}$), complicates things somewhat. Note, however, that for fixed numbers $a > 0$ and $b \in \mathbb{R}$, the equation is always uniformly parabolic in the following sense: We can write it in the form

$$R_u \frac{\partial u}{\partial t} = \Delta u + |\nabla u|^2 u \quad \text{in } (0,T)\times\Omega,$$

where R_y is the linear transformation given by the matrix

$$\begin{pmatrix} a & -by^3 & by^2 \\ by^3 & a & -by^1 \\ -by^2 & by^1 & a \end{pmatrix}$$

for any $y = (y^1, y^2, y^3) \in \mathbb{S}^2$. We have

$$a|X|^2 \le \langle X, R_y X\rangle \quad \text{and} \quad |R_y X| \le \sqrt{a^2+b^2}|X|$$

for any $y \in \mathbb{S}^2$ and any $X \in \mathbb{R}^n$.

We assume in the following that $R : N \to \mathbb{R}^{n\times n}$ is a fixed smooth function satisfying

$$|X|^2 \le c_0 \langle X, R(y)X\rangle \quad \text{and} \quad |R(y)X| \le c_0|X|$$

for a given constant c_0 and for all $y \in N$ and $X \in \mathbb{R}^n$. Then we have the following parabolic version of Lemma 3.5.

Lemma 5.5 *For every $\delta > 0$ and every $c > 0$ there exist two constants $\epsilon > 0$ and $\theta \in (0,1)$, depending only on m, N, R, δ, and c, with the following property: Suppose $u \in H^1(P_r(z_0); N)$ is a map with $\Delta u \in L^1(P_r(z_0); \mathbb{R}^n)$ such that*

$$\left| R(u)\frac{\partial u}{\partial t} - \Delta u \right| \le c|\nabla u|^2 \quad \textit{in } P_r(z_0)$$

and

$$r^{-m}\int_{P_r(z_0)} \left(|\nabla u|^2 + r^2 \left|\frac{\partial u}{\partial t}\right|^2 \right) dz \le \epsilon.$$

Then

$$\fint_{P_{\theta r}(z_0)} |u - \bar{u}_{P_{\theta r}(z_0)}|^2 \, dz \leq \theta \delta r^{-m} \int_{P_r(z_0)} \left(|\nabla u|^2 + r^2 \left| \frac{\partial u}{\partial t} \right|^2 \right) dz.$$

Proof. The arguments are similar as in the proof of Lemma 3.5. First we can assume without loss of generality that $z_0 = 0$ and $r = 1$.

Suppose the claim were false for some given $\delta > 0$ and $c > 0$. Then for any fixed $\theta \in (0, \frac{1}{2}]$ there would exist a sequence of maps $u_i \in H^1(P_1(0); N)$ such that

$$\left| R(u_i) \frac{\partial u_i}{\partial t} - \Delta u_i \right| \leq c |\nabla u_i|^2 \quad \text{in } P_1(0)$$

and

$$\int_{P_1(0)} \left(|\nabla u_i|^2 + \left| \frac{\partial u_i}{\partial t} \right|^2 \right) dz = \epsilon_i \to 0$$

as $i \to \infty$, but

$$\fint_{P_\theta(0)} |u_i - \bar{u}_i|^2 \, dz > \theta \delta \epsilon_i,$$

where

$$\bar{u}_i = \fint_{P_\theta(0)} u_i \, dz.$$

The sequence consisting of the maps

$$v_i = \frac{1}{\sqrt{\epsilon_i}} (u_i - \bar{u}_i)$$

has a subsequence that converges weakly in $H^1(P_1(0); \mathbb{R}^n)$ and strongly in $L^2(P_1(0); \mathbb{R}^n)$ to some $v \in H^1(P_1(0); \mathbb{R}^n)$. Moreover, we may assume that u_i converges strongly in $L^2(P_1(0); \mathbb{R}^n)$ to some constant $u_0 \in N$. It is now easy to show that v satisfies

$$R(u_0) \frac{\partial v}{\partial t} - \Delta v = 0 \quad \text{in } P_1(0)$$

by the same arguments as in Lemma 3.5. We also have

$$\int_{P_1(0)} \left(|\nabla v|^2 + \left| \frac{\partial v}{\partial t} \right|^2 \right) dz \leq 1$$

and

$$\fint_{P_\theta(0)} |v|^2\,dz \geq \delta\theta$$

by the strong convergence.

The parabolic equivalents of the estimates used in Lemma 3.5, on the other hand, give

$$\fint_{P_\theta(0)} |v|^2\,dz \leq C_1\theta^2$$

for a constant C_1 that depends only on m, N, and c_0. This contradicts the previous inequality if $\theta < \delta/C_1$. □

Proposition 5.1 *Let* $u \in H^1((0,T)\times\Omega; N) \cap C^0((0,T)\times\Omega; N)$ *be a solution of the equation*

$$R(u)\frac{\partial u}{\partial t} = \Delta u + A(u)(\nabla u, \nabla u) \quad \text{in } (0,T)\times\Omega. \tag{5.8}$$

Suppose there exists a number c *such that*

$$r^2 \int_{P_{r/2}(z_0)} \left|\frac{\partial u}{\partial t}\right|^2 dz \leq c \int_{P_r(z_0)} |\nabla u|^2\,dz$$

for every $P_r(z_0) \subset (0,T)\times\Omega$. *Then for every* $z_0 = (t_0, x_0) \in (0,T)\times\Omega$ *there exist a number* $\gamma > 0$ *and a radius* $r_0 > 0$ *with* $P_{2r_0}(z_0) \subset (0,T)\times\Omega$, *such that*

$$\sup_{z_1\in P_{r_0}(z_0)}\ \sup_{0<r\leq r_0} \left(r^{-m-1-\gamma} \int_{P_r(z_1)} \left(|\nabla u|^2 + r^2 \left|\frac{\partial u}{\partial t}\right|^2 \right) dz \right) < \infty. \tag{5.9}$$

Proof. Suppose $\lambda > 0$ is a fixed, but arbitrary number. We can choose $r_1 > 0$ so small that $P_{3r_1}(z_0) \subset (0,T)\times\Omega$ and

$$\operatorname*{osc}_{P_{3r_1}(z_0)} u \leq \lambda.$$

If $z_1 = (t_1, x_1) \in P_{r_1}(z_0)$ and $r \in (0, r_1]$, we can choose a cut-off function $\xi \in C_0^\infty(P_{2r}(z_1))$ with $0 \leq \xi \leq 1$, such that $\xi \equiv 1$ in $P_r(z_1)$ and $|\nabla\xi|^2 +$

$|\frac{\partial \xi}{\partial t}| \le C/r^2$. Then we have

$$\begin{aligned}\int_{P_{2r}(z_1)} \xi^2 |\nabla u|^2 \, dz &= \int_{P_{2r}(z_1)} \xi^2 \left\langle u - \bar{u}_{P_r(z_1)}, A(u)(\nabla u, \nabla u)\right\rangle dz \\ &\quad - \int_{P_{2r}(z_1)} \xi^2 \left\langle u - \bar{u}_{P_r(z_1)}, R(u)\frac{\partial u}{\partial t}\right\rangle dz \\ &\quad - 2\int_{P_{2r}(z_1)} \xi \nabla \xi \left\langle u - \bar{u}_{P_r(z_1)}, \nabla u\right\rangle dz \\ &\le C_1 \lambda \left(\int_{P_{2r}(z_1)} \xi^2 \left(|\nabla u|^2 + r^2 \left|\frac{\partial u}{\partial t}\right|^2 \right) dz + r^m \right)\end{aligned}$$

for a constant C_1 that depends only on N and c_0. If λ is chosen sufficiently small, it follows that

$$\begin{aligned}&\left(\frac{r}{2}\right)^{-m} \int_{P_{r/2}(z_1)} \left(|\nabla u|^2 + \frac{r^2}{4} \left|\frac{\partial u}{\partial t}\right|^2 \right) dz \\ &\qquad \le \frac{1}{2}(2r)^{-m} \int_{P_{2r}(z_1)} \left(|\nabla u|^2 + 4r^2 \left|\frac{\partial u}{\partial t}\right|^2 \right) dz + \frac{\epsilon}{2}\end{aligned}$$

for the constant $\epsilon > 0$ from Lemma 5.5 (belonging to some fixed, but yet undetermined $\delta > 0$). A repeated application of the last inequality implies that there exists a number $r_0 \le r_1$ such that

$$r^{-m} \int_{P_r(z_1)} \left(|\nabla u|^2 + r^2 \left|\frac{\partial u}{\partial t}\right|^2 \right) dz \le \epsilon$$

for every $r \le r_0$ and every $z_1 \in P_{r_0}(z_0)$. In particular we can now apply Lemma 5.5 for each such parabolic cylinder. Thus we have

$$⨍_{P_{\theta r}(z_1)} |u - \bar{u}_{P_{\theta r}(z_1)}|^2 \, dz \le \delta \theta r^{-m} \int_{P_r(z_1)} \left(|\nabla u|^2 + r^2 \left|\frac{\partial u}{\partial t}\right|^2 \right) dz$$

for a certain number $\theta \in (0, 1)$.

Let now $\eta \in C_0^\infty(P_{\theta r}(z_1))$ be another cut-off function with $0 \le \eta \le 1$, such that $\eta \equiv 1$ in $P_{\theta r/2}(z_1)$ and $|\nabla \eta|^2 + |\frac{\partial \eta}{\partial t}| \le C\theta^{-2} r^{-2}$. The same kind of estimate as above now gives

$$\int_{P_{\theta r}(z_1)} \eta^2 |\nabla u|^2 \, dz \leq (C_2\lambda + \delta_1) \int_{P_{\theta r}(z_1)} \left(|\nabla u|^2 + \theta^2 r^2 \left| \frac{\partial u}{\partial t} \right|^2 \right) dz$$
$$+ \frac{C_2}{\delta_1 \theta^2 r^2} \int_{P_{\theta r}(z_1)} |u - \bar{u}_{P_{\theta r}(z_1)}|^2 \, dz$$

for any $\delta_1 > 0$, where C_2 is a constant that depends only on m, N, and c_0. It follows that

$$\left(\frac{\theta r}{2}\right)^{-m-1} \int_{P_{\theta r/2}(z_1)} \left(|\nabla u|^2 + \frac{\theta^2 r^2}{4} \left| \frac{\partial u}{\partial t} \right|^2 \right) dz$$
$$\leq C_3(\lambda + \delta_1)(\theta r)^{-m-1} \int_{P_{\theta r}(z_1)} \left(|\nabla u|^2 + \theta^2 r^2 \left| \frac{\partial u}{\partial t} \right|^2 \right) dz$$
$$+ \frac{C_3 \delta}{\delta_1} r^{-m-1} \int_{P_r(z_1)} \left(|\nabla u|^2 + r^2 \left| \frac{\partial u}{\partial t} \right|^2 \right) dz$$

for a constant C_3 that depends only on m, N, and c_0. If λ, δ_1, and δ are chosen such that

$$C_3(\lambda + \delta_1) \leq \frac{1}{4} \quad \text{and} \quad \frac{C_3 \delta}{\delta_1} \leq \frac{1}{4},$$

then we can conclude that there exists a number $\gamma > 0$ such that

$$\sup_{0 < r \leq r_0} \left(r^{-m-1-\gamma} \int_{P_r(z_1)} \left(|\nabla u|^2 + r^2 \left| \frac{\partial u}{\partial t} \right|^2 \right) dz \right)$$

is bounded by a constant that is independent of z_1. Thus we have proved the claim. □

Remark 5.2 *A map $u \in H^1(P_{2r_0}(z_0); N)$ satisfying (5.9) is Hölder continuous in $P_{r_0}(z_0)$ by the parabolic version of Morrey's decay lemma. Thus we have already proved that a continuous solution of the harmonic map heat flow or the Landau-Lifshitz equation is in fact locally Hölder continuous.*

Proposition 5.2 *Let $u \in H^1(P_{2r_0}(z_0); N)$ be a solution of (5.8) satisfying (5.9). Then $\nabla u \in L^\infty(P_{r_0/2}(z_0); \mathbb{R}^{m \times n})$.*

Proof. We have

$$\sup_{z\in B_{r_0}(z_0)} \sup_{0<r\le r_0} \left(r^{-3/2-m-\gamma/2} \|\nabla u\|_{L^1(P_r(z))} + r^{-m-1-\gamma} \left\| R(u)\frac{\partial u}{\partial t} - \Delta u \right\|_{L^1(P_r(z))} \right) < \infty.$$

Thus if $\eta \in C_0^\infty(P_{r_0}(z_0))$ is a cut-off function with $\eta \equiv 1$ in $P_{r_0/2}(z_0)$, the product $v = \eta u$ satisfies

$$\sup_{z\in\mathbb{R}^{m+1}} \sup_{r>0} \left(r^{-m-1-\gamma'} \left\| R(u)\frac{\partial v}{\partial t} - \Delta v \right\|_{L^1(P_r(z))} \right) < \infty \tag{5.10}$$

for some $\gamma' > 0$.

Assume first that R_y is the identity matrix for each $y \in N$. (We are in this situation when we consider the harmonic map heat flow.) Then v can be represented using the fundamental solution of the heat equation,

$$G(t,x) = \begin{cases} (4\pi t)^{-m/2} e^{-\frac{|x|^2}{4t}}, & x \in \mathbb{R}^m,\ t > 0, \\ 0, & x \in \mathbb{R}^m,\ t \le 0. \end{cases}$$

Namely, we have

$$v(z_1) = \int_{\mathbb{R}\times\mathbb{R}^m} G(z_1 - z) \left(\frac{\partial v}{\partial t}(z) - \Delta v(z) \right) dz$$

for every $z_1 = (t_1, x_1) \in \mathbb{R} \times \mathbb{R}^m$. Thus

$$\nabla v(z_1) = \int_{\mathbb{R}\times\mathbb{R}^m} \nabla G(z_1 - z) \left(\frac{\partial v}{\partial t}(z) - \Delta v(z) \right) dz \tag{5.11}$$

almost everywhere. Because we have

$$|\nabla G| \le C r^{-m-1} \quad \text{on } P_r(0) \backslash P_{r/2}(0),$$

inequality (5.10) implies that the right-hand side of (5.11) is uniformly bounded. The arguments here are essentially the same as in the proof of Proposition 3.2. We do not repeat them in detail.

In general, we have to use a fundamental solution for the parabolic system

$$R(u)\frac{\partial v}{\partial t} - \Delta v = 0$$

instead of G. This system is uniformly parabolic with Hölder continuous coefficients. A construction of a fundamental solution for such a system is described, e. g., in Chapter 9 of A. Friedman [1964]. (The construction is too involved to give the details here.) According to these results, there exists a function $\tilde{G} : \mathbb{R}^{1+m} \times \mathbb{R}^{1+m} \to \mathbb{R}^{n \times n}$ with

$$|\nabla_x \tilde{G}(t, x, \tau, \xi)| \leq \frac{C_2}{|t-\tau|^{(m+1)/2}} e^{-C_3 \frac{|x-\xi|^2}{|t-\tau|}}$$

for certain constants $C_2, C_3 > 0$, such that

$$v(z_1) = \int_{\mathbb{R} \times \mathbb{R}^m} \tilde{G}(z_1, z) \left(R(u) \frac{\partial v}{\partial t}(z) - \Delta v(z) \right) dz$$

for every $z_1 \in \mathbb{R} \times \mathbb{R}^m$. (Here ∇_x indicates differentiation with respect to the variable x.) With the same arguments as above, the desired inequality now follows also in this situation. □

Theorem 5.1 *Suppose $u \in H^1((0,T) \times \Omega; N) \cap C^0((0,T) \times \Omega; N)$ is a solution of the harmonic map heat flow. Then $u \in C^\infty((0,T) \times \Omega; N)$.*

Proof. Like in the proof of Theorem 3.1, we obtain

$$u \in W^{1,\infty}_{\text{loc}}((0,T) \times \Omega; N)$$

from Proposition 5.1 and Proposition 5.2. Standard L^p-estimates for parabolic systems then imply

$$\nabla^2 u \in L^p_{\text{loc}}((0,T) \times \Omega; \mathbb{R}^{m \times m \times n})$$

and

$$\frac{\partial u}{\partial t} \in L^p_{\text{loc}}((0,T) \times \Omega; \mathbb{R}^n)$$

for every $p < \infty$. Higher regularity follows inductively by differentiating the system and applying standard estimates for parabolic systems with each step. □

Theorem 5.2 *Suppose $u \in H^1((0,T) \times \Omega; \mathbb{S}^2) \cap C^0((0,T) \times \Omega; \mathbb{S}^2)$ is a solution of the Landau-Lifshitz equation. Then $u \in C^\infty((0,T) \times \Omega; \mathbb{S}^2)$.*

Proof. The arguments are the same as in the previous proof. We omit the details. □

5.3 The Heat Flow into Spheres

We now consider the harmonic map heat flow for the target manifold $\mathbb{S}^{n-1} \subset \mathbb{R}^n$. Similarly as for harmonic maps, we can use arguments here which do not work for general target manifolds. In fact, for the harmonic map heat flow, not only is the reasoning easier, but we obtain better results if the target is a sphere.

The arguments we use in this section are variants of the methods from Section 3.5. They have first been used for the harmonic map heat flow by M. Feldman [1994] and by Y. Chen, J. Li, and F.-H. Lin [1995].

Let $u \in H^1((0,T) \times \Omega; \mathbb{S}^{n-1})$ be a weak solution of the harmonic map heat flow

$$\frac{\partial u}{\partial t} - \Delta u = |\nabla u|^2 u \quad \text{in } (0,T) \times \Omega. \tag{5.12}$$

We assume that u satisfies the stability condition. In particular the localized monotonicity inequality (5.6) and the other estimates derived in Section 5.1 hold true.

For harmonic maps into spheres we have used the fact that equation (3.16) can be written in the form (3.17). Similarly, we take the exterior product with u on both sides of (5.12) and obtain the equation

$$u \wedge \frac{\partial u}{\partial t} - \operatorname{div}(u \wedge \nabla u) = 0 \quad \text{in } (0,T) \times \Omega. \tag{5.13}$$

The same computations as in the proof of Lemma 3.7 show that the two equations are in fact equivalent for maps in $H^1((0,T) \times \Omega; \mathbb{S}^{n-1})$. As in the case of harmonic maps, we will use the special structure of (5.13) to estimate the norm of certain expressions in the Hardy space $\mathcal{H}^1(\mathbb{R}^m)$. This time, however, the vector field $u \wedge \nabla u$ does not have vanishing divergence. Therefore we need another version of Theorem 2.4.

Theorem 5.3 *Suppose $p > 1$. There exists a constant C_0, depending only on m and p, such that the following holds. Suppose $v \in W_0^{1,p}(B_1(0))$ and $X \in L^{p/(p-1)}(B_1(0); \mathbb{R}^m)$ with $\operatorname{div} X \in L^{p/(p-1)}(B_1(0))$. Set*

$$a = \fint_{B_1(0)} \nabla v \cdot X \, dx.$$

Then the function

$$w(x) = \begin{cases} \nabla v(x) \cdot X(x) - a & \text{if } x \in B_1(0), \\ 0 & \text{if } x \notin B_1(0), \end{cases}$$

is in $\mathcal{H}^1(\mathbb{R}^m)$ *and satisfies*

$$\|w\|_{\mathcal{H}^1(\mathbb{R}^m)} \le C_0 \|\nabla v\|_{L^p(B_1(0))} \left(\|X\|_{L^{p/(p-1)}(B_1(0))} + \|\operatorname{div} X\|_{L^{p/(p-1)}(B_1(0))} \right).$$

Proof. By Hölder's inequality we have

$$|a| \le C \|\nabla v\|_{L^p(B_1(0))} \|X\|_{L^{p/(p-1)}(B_1(0))}.$$

Let $\phi \in C_0^\infty(B_1(0))$ with $|\nabla \phi| \le 1$. For $s > 0$, set $\phi_s(x) = s^{-m}\phi(x/s)$. If $s \ge 1$, we have

$$\begin{aligned}(\phi_s * w)(x) &= \int_{B_s(x)} (\phi_s(x-y) - \phi_s(x)) w(y)\, dy \\ &\le C(|x|+1)^{-m-1} \|\nabla v\|_{L^p(B_1(0))} \|X\|_{L^{p/(p-1)}(B_1(0))}\end{aligned}$$

for any $x \in \mathbb{R}^m$ by the same arguments as in the proof of Lemma 4.7. Thus it suffices to show that

$$\int_{\mathbb{R}^m} \sup_{0<s\le 1} |\phi_s * w|\, dx \le C_1 \|\nabla v\|_{L^p(B_1(0))} \left(\|X\|_{L^{p/(p-1)}(B_1(0))} + \|\operatorname{div} X\|_{L^{p/(p-1)}(B_1(0))} \right)$$

for a constant C_1 that depends only on m and p. To this end, we proceed like in the proof of Theorem 2.4. Indeed the estimates are practically the same, except that we have now an additional term in (2.11) of the form

$$-\int_{B_s(x)} \phi_s(x-y) \operatorname{div} X(y)(v(y) - \bar{v}_{B_s(x)})\, dy.$$

This term is estimated by

$$C_2 \left(\fint_{B_s(x)} |\nabla v|^{\frac{mq}{m+q}}\, dx \right)^{\frac{m+q}{mq}} \left(\fint_{B_s(x)} |\operatorname{div} X|^{\frac{q}{q-1}}\, dx \right)^{\frac{q-1}{q}}$$

by the Hölder and the Sobolev inequality, whenever $s \le 1$. Here q is the same number as in the proof of Theorem 2.4, and C_2 is a constant that depends only on m and q. The proof is completed by the arguments in the proof of Theorem 2.4. □

Lemma 5.6 *There exists a constant* C*, depending only on* m *and* n*, with the following property. Let* $u \in H^1((0,T)\times\Omega;\mathbb{S}^{n-1})$ *be a solution of (5.12).*

Suppose $z_0 = (t_0, x_0)$, $r > 0$, and $\epsilon > 0$ are such that $P_r(z_0) \subset (0,T) \times \Omega$ and

$$[u(t,\,\cdot\,)]_{\mathrm{BMO}(B_r(x_0))} \leq \epsilon$$

for almost every $t \in (t_0 - r^2, t_0 + r^2)$. Then for any $\delta > 0$,

$$\left(\frac{r}{2}\right)^{-m} \int_{P_{r/2}(z_0)} |\nabla u|^2 \, dz \leq (C\epsilon + \delta) r^{-m} \int_{P_r(z_0)} |\nabla u|^2 \, dz + \frac{C}{\delta} r^{-m-2} \int_{P_r(z_0)} |u - \bar{u}_{P_r(z_0)}|^2 \, dz.$$

Proof. We choose a cut-off function $\eta \in C_0^\infty(P_r(z_0))$ satisfying $0 \leq \eta \leq 1$, $\eta \equiv 1$ in $P_{r/2}(z_0)$, and $|\nabla \eta|^2 + |\frac{\partial \eta}{\partial t}| \leq Cr^{-2}$. Then we have

$$\begin{aligned}
\int_{P_{r/2}(z_0)} |\nabla u|^2 \, dx &\leq \int_{P_r(z_0)} \eta |u \wedge \nabla u|^2 \, dz \\
&= \int_{P_r(z_0)} \eta \left\langle u \wedge \nabla (u - \bar{u}_{B_r(x_0)}), u \wedge \nabla u \right\rangle dz \\
&= \int_{P_r(z_0)} \eta \left\langle (u - \bar{u}_{B_r(x_0)}) \wedge \nabla u, u \wedge \nabla u \right\rangle dz \\
&\quad + \int_{P_r(z_0)} \nabla \eta \cdot \left\langle (u - \bar{u}_{B_r(x_0)}) \wedge u, u \wedge \nabla u \right\rangle dz \\
&\quad + \int_{P_r(z_0)} \eta \left\langle (u - \bar{u}_{B_r(x_0)}) \wedge u, u \wedge \frac{\partial u}{\partial t} \right\rangle dz. \qquad (5.14)
\end{aligned}$$

Here we write $\bar{u}_{B_r(x_0)}$ for the function that assigns to t the average of $u(t,\,\cdot\,)$ over $B_r(x_0)$, in contrast to $\bar{u}_{P_r(z_0)}$, which is the average of u over the parabolic cylinder $P_r(z_0)$. By $\langle\,\cdot\,,\,\cdot\,\rangle$ we denote both the scalar product in $\mathbb{R}^m \otimes \Lambda_2 \mathbb{R}^n$ and in $\Lambda_2 \mathbb{R}^n$.

For almost every fixed t we can apply (a rescaled version of) Theorem 5.3 to $X = u^i(t,\,\cdot\,)\nabla u^j(t,\,\cdot\,) - u^j(t,\,\cdot\,)\nabla u^i(t,\,\cdot\,)$ and to a function v such that $\nabla v = \nabla u^l(t,\,\cdot\,)$ in $B_r(x_0)$. Combining the result with Theorem 2.3 and Lemma 2.6, and integrating over t, we obtain

$$\begin{aligned}
&\int_{P_r(z_0)} \eta \left\langle (u - \bar{u}_{B_r(x_0)}) \wedge \nabla u, u \wedge \nabla u \right\rangle dz \\
&\qquad \leq C\epsilon \|\nabla u\|_{L^2(P_r(z_0))} \left(\|\nabla u\|_{P_r(z_0))} + r \left\| \frac{\partial u}{\partial t} \right\|_{L^2(P_r(z_0))} \right). \qquad (5.15)
\end{aligned}$$

The Hölder inequality implies that

$$\begin{aligned}
&\int_{P_r(z_0)} \nabla\eta \left\langle (u - \bar{u}_{B_r(x_0)}) \wedge u, u \wedge \nabla u \right\rangle dz \\
&\le Cr^{-1} \left(\int_{t_0-r^2}^{t_0+r^2} \int_{B_r(x_0)} |u - \bar{u}_{B_r(x_0)}|^2 \, dx \, dt \right)^{1/2} \|\nabla u\|_{L^2(P_r(z_0))} \\
&\le Cr^{-1} \|u - \bar{u}_{P_r(z_0)}\|_{L^2(P_r(z_0))} \|\nabla u\|_{L^2(P_r(z_0))} \qquad (5.16)
\end{aligned}$$

(because

$$\int_{B_r(x_0)} |u(t,x) - \bar{u}_{B_r(x_0)}(t)|^2 \, dx \le \int_{B_r(x_0)} |u - \bar{u}_{P_r(z_0)}|^2 \, dx$$

for almost every t), and

$$\begin{aligned}
&\int_{P_r(z_0)} \eta \left\langle (u - \bar{u}_{B_r(x_0)}) \wedge u, u \wedge \frac{\partial u}{\partial t} \right\rangle dz \\
&\qquad \le C \|u - \bar{u}_{P_r(z_0)}\|_{L^2(P_r(z_0))} \left\| \frac{\partial u}{\partial t} \right\|_{L^2(P_r(z_0))}. \qquad (5.17)
\end{aligned}$$

To complete the proof, we simply have to combine (5.14)–(5.17) with Young's inequality. □

Lemma 5.7 *There exist $\epsilon_0 > 0$ and $\theta \in (0,1)$, both depending only on m and n, with the following property. Suppose $u \in H^1((0,T) \times \Omega; \mathbb{S}^{n-1})$ is a solution of (5.12) satisfying the stability condition. Then for every parabolic cylinder $P_r(z_0) \subset (0,T) \times \Omega$ with*

$$r^{-m} \int_{P_r(z_0)} |\nabla u|^2 \, dz \le \epsilon_0,$$

the inequality

$$(\theta r)^{-m} \int_{P_{\theta r}(z_0)} \left(|\nabla u|^2 + \theta^2 r^2 \left| \frac{\partial u}{\partial t} \right|^2 \right) dz \le \frac{1}{2} r^{-m} \int_{P_r(z_0)} |\nabla u|^2 \, dz$$

holds.

Proof. Lemma 5.4 implies that

$$\sup_{t_0 - r^2/4 < t < t_0 + r^2/4} [u(t, \cdot)]_{\mathrm{BMO}(B_{r/2}(x_0))} \le C\epsilon_0.$$

Thus Lemma 5.6 yields

$$(\theta r)^{-m} \int_{P_{\theta r}(z_0)} |\nabla u|^2 \, dz \leq (C\epsilon_0 + \delta)(2\theta r)^{-m} \int_{P_{2\theta r}(z_0)} |\nabla u|^2 \, dz + \frac{C}{\delta}(2\theta r)^{-m-2} \int_{P_{2\theta r}(z_0)} |u - \bar{u}_{P_{2\theta r}(z_0)}|^2 \, dz$$

for any $\delta > 0$, provided that $\theta \leq \frac{1}{4}$. The last term can be estimated further with Lemma 5.5, provided that ϵ_0 is sufficiently small. If δ is also small enough, this gives the desired estimate for the integral of $|\nabla u|^2$. To show the estimate for the integral of $|\frac{\partial u}{\partial t}|^2$, we apply Lemma 5.2. □

Proposition 5.3 *There exists a number $\epsilon > 0$, depending only on m and n, such that every solution $u \in H^1((0,T) \times \Omega; \mathbb{S}^{n-1})$ of (5.12) satisfying the stability condition and*

$$r^{-m} \int_{P_r(z_0)} |\nabla u|^2 \, dz \leq \epsilon$$

for some parabolic cylinder $P_r(z_0) \subset (0,T) \times \Omega$, is smooth in $P_{r/2}(z_0)$; i. e., $u \in C^\infty(P_{r/2}(z_0); \mathbb{S}^{n-1})$.

Proof. The arguments here are very similar to the ones for the elliptic problems. If ϵ is sufficiently small, we can apply Lemma 5.7 repeatedly, and we obtain

$$s^{-m} \int_{P_s(z_1)} \left(|\nabla u|^2 + s^2 \left| \frac{\partial u}{\partial t} \right| \right) dz \leq C\epsilon s^\alpha$$

for all $z_1 \in P_{r/2}(z_0)$ and $0 < s \leq r/2$, where α is a positive number. The parabolic version of Morrey's decay lemma (Lemma 2.2) implies Hölder continuity of u in $P_{r/2}(z_0)$. Then Theorem 5.1 implies that u is smooth in $P_{r/2}(z_0)$. □

Theorem 5.4 *Let $u \in H^1((0,T) \times \Omega; \mathbb{S}^{n-1})$ be a solution of (5.12) satisfying the stability condition. There exists a relatively closed set $\Sigma \subset (0,T) \times \Omega$ of vanishing m-dimensional parabolic Hausdorff measure such that $u \in C^\infty(((0,T) \times \Omega) \backslash \Sigma; \mathbb{S}^{n-1})$.*

Proof. Again the arguments are basically the same as in the proof of Theorem 3.2. Instead of balls, however, we now use parabolic cylinders. Accordingly, we obtain a statement about the parabolic Hausdorff measure of the singular set

$$\Sigma = \left\{ z_0 \in (0,T) \times \Omega : \liminf_{r \searrow 0} \left(r^{-m} \int_{P_r(z_0)} |\nabla u|^2 \, dz \right) \geq \epsilon \right\},$$

where $\epsilon > 0$ is the number from Proposition 5.3. We leave it to the reader to make the necessary adaptions in the arguments. □

5.4 The Heat Flow into General Target Manifolds

We now replace the target manifold $\mathbb{S}^{n-1}$ considered in the previous section by a general smooth, compact Riemannian manifold $N \subset \mathbb{R}^n$. The harmonic map heat flow is then given by the equation

$$\frac{\partial u}{\partial t} - \Delta u = A(u)(\nabla u, \nabla u) \quad \text{in } (0,T) \times \Omega. \tag{5.18}$$

Let $u \in H^1((0,T) \times \Omega; N)$ be a solution of (5.18) satisfying the stability condition. We want to prove a partial regularity result for u. Unfortunately, this is difficult in general. Unlike the methods for harmonic maps into spheres, the arguments that prove partial regularity for weakly harmonic maps into general target manifolds (i. e., the arguments of Section 3.7) are not so easily adapted to a parabolic equation. The main reason for this difficulty is that we were forced to work with the L^1-norm of ∇u in Section 3.7 rather than the L^2-norm (which would be natural from the variational point of view), in order to overcome some technical problems.

Thus we need another approach. One possibility is to fix t and regard $u(t, \cdot)$ as an almost harmonic map. Then we can try to use the methods of Chapter 4. It turns out that this approach works as long as $m \leq 4$, but it fails for $m \geq 5$. The dimension 4 is a natural limit for this method. We have

$$\frac{\partial u}{\partial t}(t, \cdot) \in L^2_{\text{loc}}(\Omega; \mathbb{R}^n)$$

for almost every t by the local energy inequality, but we do not have any L^p-estimates for $\frac{\partial u}{\partial t}$ with $p > 2$. Thus in order to apply any of the results of Chapter 4, we must have $m \leq 4$. The restriction to small dimensions is imposed by our choice of the method, not by some intrinsic property of the problem; but it is unclear how similar results for large dimensions can be obtained. The problem is open for $m \geq 5$.

Assume now that $m \leq 4$. Remember that the results of Section 5.1 are independent of the target manifold. Thus we still have the local monotonicity formula given by Lemma 5.3 and the estimate from Lemma 5.4, as well as the local energy inequality (5.5). Together with Proposition 4.2, this implies the following.

Lemma 5.8 *Suppose $m \leq 4$. There exist two constants $\epsilon > 0$ and C, depending only on m and N, such that the following holds. Let $u \in H^1(P_r(z_0); N)$ be a solution of (5.18) satisfying the stability condition. If*

$$r^{-m} \int_{P_r(z_0)} |\nabla u|^2 \, dz \leq \epsilon,$$

then

$$\begin{aligned}\left(\frac{r}{2}\right)^{-m} \int_{P_{r/2}(z_0)} & \left(|\nabla u|^2 + \frac{r^2}{4}\left|\frac{\partial u}{\partial t}\right|^2\right) dz \\ & \leq \frac{1}{2} r^{-m} \int_{P_r(z_0)} \left(|\nabla u|^2 + r^2 \left|\frac{\partial u}{\partial t}\right|^2\right) dz \\ & \quad + C r^{-m-2} \int_{P_r(z_0)} |u - \bar{u}_{P_r(z_0)}|^2 \, dz. \qquad (5.19)\end{aligned}$$

Proof. If ϵ is sufficiently small, then Lemma 5.4 implies that $u(t, \cdot)$ satisfies the conditions of Proposition 4.2 for almost every $t \in (t_0 - r^2/4, t_0 + r^2/4)$ (where $z_0 = (t_0, x_0)$). Integrating the inequality obtained from Proposition 4.2 over the interval $(t_0 - r^2/256, t_0 + r^2/256)$ and using Lemma 5.2, we find

$$\begin{aligned}\left(\frac{r}{16}\right)^{-m} \int_{P_{r/16}(z_0)} & \left(|\nabla u|^2 + \frac{r^2}{256}\left|\frac{\partial u}{\partial t}\right|^2\right) dz \\ & \leq \delta r^{-m} \int_{P_r(z_0)} \left(|\nabla u|^2 + r^2 \left|\frac{\partial u}{\partial t}\right|^2\right) dz \\ & \quad + \frac{C}{\delta} r^{-m-2} \int_{P_r(z_0)} |u - \bar{u}_{P_r(z_0)}|^2 \, dz\end{aligned}$$

for an arbitrarily small $\delta > 0$. The proof can then be finished with a simple covering argument. □

Lemma 5.9 *Suppose $m \leq 4$. There exist $\epsilon_0 > 0$ and $\theta \in (0, 1)$, both depending only on N, with the following property. Suppose $u \in H^1(P_r(z_0); N)$*

is a solution of (5.18) satisfying the stability condition. If

$$r^{-m} \int_{P_r(z_0)} |\nabla u|^2 \, dz \leq \epsilon_0,$$

then the inequality

$$(\theta r)^{-m} \int_{P_{\theta r}(z_0)} \left(|\nabla u|^2 + \theta^2 r^2 \left| \frac{\partial u}{\partial t} \right|^2 \right) dz \leq \frac{1}{2} r^{-m} \int_{P_r(z_0)} |\nabla u|^2 \, dz$$

is satisfied.

Proof. This follows from Lemma 5.8, Lemma 5.5, and Lemma 5.2. □

We can continue to use arguments we have seen before. Applying Lemma 5.9 repeatedly and using Lemma 2.2 and Theorem 5.1 gives the result of the next proposition. The theorem following it is then proved by the usual covering arguments (which are described in the proofs of Theorem 3.2 and Theorem 5.4).

Proposition 5.4 *Suppose $m \leq 4$. There exists a number $\epsilon > 0$, depending only on N, such that every solution $u \in H^1(P_r(z_0); N)$ of (5.18) satisfying the stability condition and*

$$r^{-m} \int_{P_r(z_0)} |\nabla u|^2 \, dz \leq \epsilon,$$

is in the space $C^\infty(P_{r/2}(z_0); N)$.

Theorem 5.5 *For $m \leq 4$, let $u \in H^1((0,T) \times \Omega; N)$ be a solution of (5.18) satisfying the stability condition. There exists a relatively closed set $\Sigma \subset (0,T) \times \Omega$ of vanishing m-dimensional parabolic Hausdorff measure such that $u \in C^\infty(((0,T) \times \Omega) \backslash \Sigma; N)$.*

We have seen in Chapter 3 that the case $m = 2$ permits particularly nice regularity results for weakly harmonic maps. The same is true for the harmonic map heat flow. For this problem, however, we cannot drop the stability condition entirely. We still need the local energy inequality (5.5). But if we have this inequality, we can use the following reasoning: Suppose $m = 2$. Let $z_0 = (t_0, x_0) \in \mathbb{R} \times \mathbb{R}^2$ and $r > 0$. Suppose $u \in H^1(P_r(z_0); N)$ is a solution of (5.18) such that

$$\int_{P_r(z_0)} |\nabla u|^2 \, dz \leq \epsilon.$$

If (5.5) is satisfied for any non-negative function $\xi \in C_0^\infty(P_r(z_0))$, it follows that

$$\int_{B_{r/2}(x_1)} |\nabla u(t_1, x)|^2 \, dx \leq C\epsilon$$

for any $t_1 \in (t_0 - r^2/4, t_0 + r^2/4)$. Since $m = 2$, this means

$$[u(t_1, \cdot\,)]_{\mathrm{BMO}(B_{r/2}(x_0))} \leq C\epsilon.$$

Hence if ϵ is sufficiently small, the arguments of this section imply that u is smooth in $P_{r/2}(z_0)$.

Theorem 5.6 *Suppose $m = 2$. If $u \in H^1((0,T)\times\Omega; N)$ is a weak solution of (5.18) satisfying (5.5) for any non-negative function $\xi \in C_0^\infty((0,T)\times\Omega)$, then there exists a finite set $\Sigma \subset (0,T) \times \Omega$ such that $u \in C^\infty(((0,T) \times \Omega)\backslash\Sigma; N)$.*

Proof. Choose a finite number of pairwise distinct points $z_1, \ldots, z_K \in (0,T) \times \Omega$ such that u is not smooth in any neighborhood of any of these points. There exists some $r > 0$, such that $P_r(z_k) \subset (0,T) \times \Omega$ for $k = 1, \ldots, K$ and

$$P_r(z_k) \cap P_r(z_l) = \emptyset \quad \text{if } k \neq l.$$

By the arguments above, we have

$$\int_{P_r(z_k)} |\nabla u|^2 \, dz \geq \epsilon$$

for $k = 1, \ldots, K$, where $\epsilon > 0$ is a constant that depends only on N. Hence

$$K \leq \frac{1}{\epsilon} \sum_{k=1}^{K} \int_{P_r(z_k)} |\nabla u|^2 \, dz \leq \frac{1}{\epsilon} \int_0^T \int_\Omega |\nabla u|^2 \, dx \, dt.$$

This clearly implies the claim of the theorem. □

Remark 5.3 *A more thorough analysis of the singularities reveals that the energy of $u(t, \cdot\,)$ must decrease at least by a fixed amount at every singular point. If a global energy inequality of the form (1.9), as mentioned in the introduction, holds, the maximal number of singularities can then be estimated by a constant times the initial energy. These arguments are due to M. Struwe [1985] (see also K.-C. Chang [1989]). We do not discuss them in detail here.*

5.5 The Landau-Lifshitz Equation

In this section we study the Landau-Lifshitz equation that arises in the theory of ferromagnetism. This is an equation for maps with values in the 2-sphere $\mathbb{S}^2 \subset \mathbb{R}^3$. Suppose $a > 0$ and $b \in \mathbb{R}$ are two given constants. We now denote by $\wedge$ the vector product in $\mathbb{R}^3$ (or, equivalently, we identify $\Lambda_2\mathbb{R}^3$ with $\mathbb{R}^3$). The equation we consider can be written in (at least) two different forms. We first consider the equation

$$\frac{\partial u}{\partial t} = -a\, u \wedge (u \wedge \Delta u) - b\, u \wedge \Delta u \quad \text{in } (0,T) \times \Omega. \tag{5.20}$$

Suppose this equation is solved for initial data in $H^1(\Omega; \mathbb{S}^2)$. We have $\frac{\partial u}{\partial t} \perp u$ and thus

$$\frac{\partial}{\partial t}|u|^2 = 0$$

almost everywhere. Hence the condition $|u| = 1$ almost everywhere remains satisfied. We assume throughout this section that we have solutions with values in the unit sphere.

For convenience we also assume that

$$a^2 + b^2 = 1.$$

If this condition is not true, we can rescale the time axis and transform (5.20) into an equation with this property. Now note the following: If $y \in \mathbb{S}^2$ and $X \in \mathbb{R}^3$, then $y \wedge X$ is the vector resulting in projecting X onto $T_y\mathbb{S}^2$ orthogonally and then rotating it by an angle of $\frac{\pi}{2}$. Thus the expression

$$-a\, u \wedge (u \wedge \Delta u) - b\, u \wedge \Delta u$$

in (5.20) corresponds to an orthogonal projection of $\Delta u(u)$ onto $T_{u(x)}\mathbb{S}^2$ followed by a rotation by the angle

$$-\arctan\frac{b}{a}$$

in $T_{u(x)}\mathbb{S}^2$ for almost every $x \in \Omega$ (provided that $\Delta u(x)$ is well-defined almost everywhere). The orthogonal projection of $\Delta u(x)$ onto $T_{u(x)}\mathbb{S}^2$ is

$$\Delta u(x) + |\nabla u(x)|^2 u(x).$$

Instead of rotating this vector by the angle $-\arctan(b/a)$, we can rotate the left-hand side in (5.20), that is, the time-derivative $\frac{\partial u}{\partial t}$, by $\arctan(b/a)$.

Thus if u is sufficiently regular (for instance, $u \in H^1(\Omega; \mathbb{S}^2)$ and $\Delta u \in L^1(\Omega; \mathbb{R}^3)$), then (5.20) is equivalent to

$$a\,\frac{\partial u}{\partial t} + b\,u \wedge \frac{\partial u}{\partial t} = \Delta u + |\nabla u|^2 u \quad \text{in } (0,T) \times \Omega. \tag{5.21}$$

In this form, the Landau-Lifshitz equation is also known as the Landau-Lifshitz-Gilbert equation. The equivalence of (5.20) and (5.21) can also be shown in a more algebraic way by taking the exterior product with u on both sides of (5.20) and using the identity

$$X \times (Y \times Z) = (X \cdot Z)Y - (X \cdot Y)Z.$$

Because (5.21) is more easily accessible to the methods from the theory of harmonic maps than (5.20), this is the equation we study in the rest of this section. Using the abbreviation

$$R_y X = a\,X + b\,y \wedge X$$

for $y \in \mathbb{S}^2$ and $X \in \mathbb{R}^3$, we can also write (5.21) in the form

$$R_u \frac{\partial u}{\partial t} = \Delta u + |\nabla u|^2 u \quad \text{in } (0,T) \times \Omega.$$

This simplifies the notation somewhat.

For the same reasons as for the harmonic map heat flow, it is natural to assume that $u \in H^1((0,T) \times \Omega; \mathbb{S}^2)$. We use a similar stability condition as in the previous sections of this chapter. For $\psi \in C_0^\infty((0,T) \times \Omega; \mathbb{R}^m)$ and $\tau \in C_0^\infty((0,T) \times \Omega; [0,\infty))$, we define for this purpose

$$u_s(t,x) = u(t + s\tau(t,x), x + s\psi(t,x)). \tag{5.22}$$

Definition 5.2 A weak solution $u \in H^1((0,T) \times \Omega; \mathbb{S}^2)$ of (5.21) is said to satisfy the *stability condition*, if

$$\int_0^T \int_\Omega \left\langle R_u \frac{\partial u}{\partial t}, \left.\frac{\partial u_s}{\partial s}\right|_{s=0} \right\rangle dx\,dt + \left.\frac{d}{ds}\right|_{s=0} \int_0^T E(u_s)\,dt \leq 0 \tag{5.23}$$

for all variations of u of the form (5.22).

The difficulties we have for this problem are of a different kind as in the previous section. Here we have the target manifold $\mathbb{S}^2$. This allows in principle to apply the arguments of Section 5.3. The main problem for the Landau-Lifshitz equation is the lack of a parabolic monotonicity formula, even under the stability condition. Nevertheless, the remedy is the same as before: We only consider domains of small dimensions, so that we can work with time slices of the form $\{t\} \times \Omega$ and use the results for almost harmonic maps from Chapter 4. Thus we assume that $2 \le m \le 4$ for the rest of this section.

Since we cannot prove an equivalent of the local monotonicity formula (5.6) for the Landau-Lifshitz equation, we need a substitute. The next lemma will give the necessary tools to derive it.

Lemma 5.10 *Let $u \in H^1((0,T) \times \Omega; \mathbb{S}^2)$ be a weak solution of (5.21) that satisfies the stability condition. Let $\psi \in C_0^\infty(\Omega; \mathbb{R}^m)$ and $\tau \in C_0^\infty(\Omega \times (0,T); [0,\infty))$. Then*

$$\int_{\{t\}\times\Omega} \left(\psi \cdot \left\langle R_u \frac{\partial u}{\partial t}, \nabla u \right\rangle - \frac{1}{2} \operatorname{div} \psi \, |\nabla u|^2 + \sum_{\alpha,\beta=1}^{m} \frac{\partial \psi^\alpha}{\partial x^\beta} \left\langle \frac{\partial u}{\partial x^\alpha}, \frac{\partial u}{\partial x^\beta} \right\rangle \right) dx = 0 \tag{5.24}$$

for almost every $t \in (0,T)$, and

$$\int_{\{t_2\}\times\Omega} \tau |\nabla u|^2 \, dx - \int_{\{t_1\}\times\Omega} \tau |\nabla u|^2 \, dx + a \int_{t_1}^{t_2} \int_\Omega \tau \left| \frac{\partial u}{\partial t} \right|^2 dx \, dt \le \int_{t_1}^{t_2} \int_\Omega \left(\frac{\partial \tau}{\partial t} |\nabla u|^2 - \nabla \tau \cdot \left\langle \nabla u, \frac{\partial u}{\partial t} \right\rangle \right) dx \, dt \tag{5.25}$$

for almost all t_1, t_2 with $0 \le t_1 \le t_2 \le T$.

Proof. Following the same calculations as in the derivation of Lemma 5.1 in Section 5.1 (but with $\tau = 0$), we see that

$$\int_0^T \int_\Omega \left(\psi \cdot \left\langle R_u \frac{\partial u}{\partial t}, \nabla u \right\rangle - \frac{1}{2} \operatorname{div} \psi |\nabla u|^2 + \sum_{\alpha=1}^{m} \nabla \psi^\alpha \cdot \left\langle \frac{\partial u}{\partial x^\alpha}, \nabla u \right\rangle \right) dx \, dt \le 0$$

for all $\psi \in C_0^\infty((0,T) \times \Omega; \mathbb{R}^m)$. Applying the same inequality for $-\psi$ instead of ψ we see that we have in fact equality. Identity (5.24) follows immediately. Inequality (5.25) is proved with the same arguments as the inequality in Remark 5.1 and the observation

$$\left\langle R_u \frac{\partial u}{\partial t}, \frac{\partial u}{\partial t} \right\rangle = a \left| \frac{\partial u}{\partial t} \right|^2 .$$

□

Remark 5.4 *Choosing $\tau(t,x) = \xi(x)$ for some $\xi \in C_0^\infty(\Omega;[0,\infty))$, we obtain a local energy inequality of the form*

$$\frac{1}{2}\int_\Omega \xi(x)(|\nabla u(t_2,x)|^2 - |\nabla u(t_1,x)|^2)\,dx + a\int_{t_1}^{t_2}\int_\Omega \xi\left|\frac{\partial u}{\partial t}\right|^2 dx\,dt$$
$$\leq -\int_{t_1}^{t_2}\int_\Omega \nabla\xi \cdot \left\langle \frac{\partial u}{\partial t}, \nabla u\right\rangle dx\,dt. \qquad (5.26)$$

With the arguments from the proof of Lemma 5.2, we can then also show that

$$r^{2-m}\int_{B_{r/2}(x_0)} |\nabla u(t,x)|^2\,dx + r^{2-m}\int_{P_{r/2}(z_0)}\left|\frac{\partial u}{\partial t}\right|^2 dz$$
$$\leq C_0 r^{-m}\int_{P_r(z_0)} |\nabla u|^2\,dz \qquad (5.27)$$

for a constant C_0 that depends only on m and a.

We also note that identity (5.24) is of the same type as the condition (4.2) that we used to prove a monotonicity formula for almost harmonic maps. Lemma 4.1 thus implies the following.

Lemma 5.11 *Suppose $m = 3$ or $m = 4$. Let $u \in H^1((0,T)\times\Omega;\mathbb{S}^2)$ be a weak solution of (5.21) satisfying the stability condition. Suppose $B_s(x_0) \subset B_r(x_0) \subset \Omega$. Set*

$$\Phi(u;t,\rho) = \rho^{2-m}\int_{\{t\}\times B_\rho(x_0)}\left(|\nabla u|^2 - \frac{2}{m-2}\left\langle (x-x_0)\cdot\nabla u, R_u\frac{\partial u}{\partial t}\right\rangle\right) dx$$

for $s \leq \rho \leq r$ and for all $t \in (0,T)$ such that this expression is well-defined (which is the case for almost every t). Then

$$\Phi(u;t,r) - \Phi(u;t,s)$$
$$= 2\int_{B_r(x_0)\setminus B_s(x_0)}\left(\frac{|(x-x_0)\cdot\nabla u|^2}{|x-x_0|^m} - \frac{\langle (x-x_0)\cdot\nabla u, R_u\frac{\partial u}{\partial t}\rangle}{(m-2)|x-x_0|^{m-2}}\right) dx$$

for almost every $t \in (0,T)$. In particular,

$$s^{2-m}\int_{B_s(x_0)} |\nabla u(t,x)|^2\,dx \leq (1+\delta)r^{2-m}\int_{B_r(x_0)} |\nabla u(t,x)|^2\,dx$$
$$+ C_0 r^{4-m}\int_{B_r(x_0)}\left|\frac{\partial u}{\partial t}(t,x)\right|^2 dx$$

for every $\delta > 0$ and almost every $t \in (0,T)$, where C_0 is a constant that depends only on a and δ.

We can now proceed with the same type of arguments that we already used in Section 5.4.

Lemma 5.12 *For any $\delta > 0$ there exist two numbers C_0 and $\epsilon_0 > 0$, depending only on m, a, and δ, such that the following holds. Suppose $u \in H^1(P_r(z_0); \mathbb{S}^2)$ is a weak solution of (5.21) satisfying the stability condition. If*

$$r^{-m} \int_{P_r(z_0)} |\nabla u|^2 \, dz \leq \epsilon_0, \tag{5.28}$$

then the inequality

$$r^{-m} \int_{P_{r/8}(z_0)} |\nabla u|^2 \, dz \leq \delta r^{-m} \int_{P_r(z_0)} |\nabla u|^2 \, dz + C_0 \fint_{P_r(z_0)} |u - \bar{u}_{P_r(z_0)}|^2 \, dz$$

holds.

Proof. All of the quantities appearing in this lemma are invariant under the transformation $(t,x) \mapsto (r^2 t + t_0, rx + x_0)$. We may thus assume that $P_r(z_0) = P_1(0)$.

Given a number $\lambda \in (0,1)$, we infer from (5.27) that there exists a set $\Lambda \subset (-\frac{1}{2}, \frac{1}{2})$ of measure $|\Lambda| \leq \lambda$, such that

$$\int_{B_{1/2}(0)} \left| \frac{\partial u}{\partial t}(t,x) \right|^2 \, dx \leq \frac{C}{a\lambda} \int_{P_1(0)} |\nabla u|^2 \, dz$$

for all $t \in (-\frac{1}{2}, \frac{1}{2}) \backslash \Lambda$. Moreover, for almost every $t \in (-\frac{1}{2}, \frac{1}{2})$, we have

$$\int_{B_{1/2}(0)} |\nabla u(t,x)|^2 \, dx \leq C \int_{P_1(0)} |\nabla u|^2 \, dz$$

by the same inequality. If $t \notin \Lambda$, we even obtain the estimate

$$\sup_{B_r(x_0) \subset B_{1/4}(0)} \left(r^{2-m} \int_{B_r(x_0)} |\nabla u(t,x)|^2 \, dx \right) \leq \frac{C}{a\lambda} \int_{P_1(0)} |\nabla u|^2 \, dz$$

from Lemma 5.11.

In the following, let t be a fixed number in $(-\frac{1}{2}, \frac{1}{2})\backslash\Lambda$. Choose $\zeta \in C_0^\infty(B_{1/4}(0))$ with $0 \le \zeta \le 1$, $\zeta \equiv 1$ in $B_{1/8}(0)$, and $|\nabla\zeta| \le 16$. Note that

$$\begin{aligned}
\int_{B_1(0)} \zeta(x)|\nabla u(t,x)|^2\, dx \\
&= -\int_{B_1(0)} \zeta(x) \left\langle u(t,x) - \bar{u}_{B_1(0)}(t), R_{u(t,x)} \frac{\partial u}{\partial t}(t,x) \right\rangle dx \\
&\quad + \int_{B_1(0)} \zeta(x) \left\langle u(t,x) - \bar{u}_{B_1(0)}(t), |\nabla u(t,x)|^2 u(t,x) \right\rangle dx \\
&\quad - \int_{B_1(0)} \nabla\zeta(x) \cdot \left\langle u(t,x) - \bar{u}_{B_1(0)}(t), \nabla u(t,x) \right\rangle dx. \qquad (5.29)
\end{aligned}$$

Here we write

$$\bar{u}_{B_1(0)}(t) = \fint_{B_1(0)} u(t,x)\, dx.$$

Since u takes values in $\mathbb{S}^2$ almost everywhere, we have

$$|\nabla u|^2 u^i = \sum_{j=1}^{3} \nabla u^j \cdot (u^i \nabla u^j - u^j \nabla u^i).$$

Furthermore,

$$\operatorname{div}(u^i \nabla u^j - u^j \nabla u^i) = u^i w^j - u^j w^i,$$

where

$$w = R_u \frac{\partial u}{\partial t}.$$

Hence

$$\begin{aligned}
&\left\| \operatorname{div}(\zeta(u^i(t,\cdot)\nabla u^j(t,\cdot) - u^j(t,\cdot)\nabla u^i(t,\cdot))) \right\|_{L^2(B_1(0))} \\
&\qquad \le C\|\nabla u(t,\cdot)\|_{L^2(B_{1/4}(0))} + C\left\| \frac{\partial u}{\partial t}(t,\cdot) \right\|_{L^2(B_{1/4}(0))} \\
&\qquad \le \frac{C}{\sqrt{a\lambda}} \|\nabla u\|_{L^2(P_1(0))}.
\end{aligned}$$

Extending the functions $u - \bar{u}_{B_1(0)}$ and ∇u appropriately to $\mathbb{R}^m$ and applying Theorem 5.3 and Theorem 2.3, we find that

$$\int_{B_1(0)} \zeta(x) \left\langle u(t,x) - \bar{u}_{B_1(0)}(t), |\nabla u(t,x)|^2 u(t,x) \right\rangle dx \\ \leq \frac{C\sqrt{\epsilon_0}}{a\lambda} \int_{P_1(0)} |\nabla u|^2 \, dz.$$

Using Hölder's and Young's inequality to estimate the other terms on the right-hand side of (5.29), we obtain

$$\int_{B_{1/8}(0)} |\nabla u(t,x)|^2 \, dx \leq \left(\frac{C\sqrt{\epsilon_0}}{a\lambda} + \frac{\delta}{2} \right) \int_{P_1(0)} |\nabla u|^2 \, dz \\ + \frac{C}{\delta} \int_{B_1(0)} |u(t,x) - \bar{u}_{B_1(0)}(t)|^2 \, dx.$$

We note that

$$\int_{B_1(0)} |u(t,x) - \bar{u}_{B_1(0)}(t)|^2 \, dx \leq \int_{B_1(0)} |u(t,x) - \bar{u}_{P_1(0)}|^2 \, dx$$

for every t. Hence

$$\int_{P_{1/8}(0)} |\nabla u|^2 \, dx \leq \left(\frac{C\sqrt{\epsilon_0}}{a\lambda} + \frac{\delta}{2} + C\lambda \right) \int_{P_1(0)} |\nabla u|^2 \, dz \\ + \frac{C}{\delta} \int_{P_1(0)} |u - \bar{u}_{P_1(0)}|^2 \, dz.$$

For the right choice of λ and ϵ_0, this implies the claim. □

The remaining arguments are exactly the same as in the previous sections. First we combine Lemma 5.12 with Lemma 5.5. We obtain the following energy decay lemma.

Lemma 5.13 *Suppose $m \leq 4$. There exist $\epsilon_0 > 0$ and $\theta \in (0,1)$, both depending only on a, with the following property. Suppose $u \in H^1(P_r(z_0); \mathbb{S}^2)$ is a solution of (5.21) satisfying the stability condition. If*

$$r^{-m} \int_{P_r(z_0)} |\nabla u|^2 \, dz \leq \epsilon_0,$$

then the inequality

$$(\theta r)^{-m} \int_{P_{\theta r}(z_0)} \left(|\nabla u|^2 + \theta^2 r^2 \left| \frac{\partial u}{\partial t} \right|^2 \right) dz \leq \frac{1}{2} r^{-m} \int_{P_r(z_0)} |\nabla u|^2 \, dz$$

is satisfied.

Combining Lemma 5.13 with Lemma 2.2, we obtain Hölder continuity of a solution of the Landau-Lifshitz equation satisfying the stability condition, provided that the energy is small in a parabolic cylinder. Theorem 5.2 then implies the next proposition. The theorem following it is proved by the usual covering arguments.

Proposition 5.5 *Suppose $m \leq 4$. There exists a number $\epsilon > 0$, depending only on a, such that every solution $u \in H^1(P_r(z_0); \mathbb{S}^2)$ of (5.21) satisfying the stability condition and*

$$r^{-m} \int_{P_r(z_0)} |\nabla u|^2 \, dz \leq \epsilon,$$

is in the space $C^\infty(P_{r/2}(z_0); \mathbb{S}^2)$.

Theorem 5.7 *For $m \leq 4$, let $u \in H^1((0,T) \times \Omega; \mathbb{S}^2)$ be a solution of (5.21) satisfying the stability condition. There exists a relatively closed set $\Sigma \subset (0,T) \times \Omega$ of vanishing m-dimensional parabolic Hausdorff measure such that $u \in C^\infty(((0,T) \times \Omega) \backslash \Sigma; \mathbb{S}^2)$.*

For two-dimensional domains, we can again apply the same reasoning as in the proof of Theorem 5.6 in order to obtain a better result. We leave it to the reader to make the obvious adaptions in the arguments.

Theorem 5.8 *Suppose $m = 2$. If $u \in H^1((0,T) \times \Omega; \mathbb{S}^2)$ is a weak solution of (5.21) satisfying the local energy inequality (5.26) for all non-negative functions $\xi \in C_0^\infty((0,T) \times \Omega)$, then there exists a finite set $\Sigma \subset (0,T) \times \Omega$ such that $u \in C^\infty(((0,T) \times \Omega) \backslash \Sigma; \mathbb{S}^2)$.*

5.6 Further Remarks

5.6.1 *The Harmonic Map Heat Flow into Homogeneous Manifolds*

The arguments of Section 5.3 can be generalized to manifolds satisfying the conditions of Section 3.9.1. To this end, the vector fields $X_1, \ldots, X_K, Y_1, \ldots, Y_K$ constructed by F. Hélein [1991b] and mentioned in Section 3.9.1 are used again. Then equation (5.1) can be written in the form

$$\left\langle \frac{\partial u}{\partial t}, X_l(u) \right\rangle = \operatorname{div} \langle \nabla u, X_l(u) \rangle \quad \text{in } (0,T) \times \Omega, \; l = 1, \ldots, K.$$

The generalization of the methods of Section 5.3 is then quite easy. We do not go into details here. The reader who is interested in a more thorough discussion can find it in a paper by Y. Chen and C. Wang [1996].

5.6.2 *The Landau-Lifshitz Equation for Almost-Kähler Manifolds*

The special structure of the sphere $\mathbb{S}^2$ is important for the Landau-Lifshitz equation not only for the purpose of proving regularity. We also used it in Section 5.5 in order to derive some of the basic properties of the equation; in fact we used properties of the space $\mathbb{R}^3$ in order to write down the Landau-Lifshitz equation. Nevertheless, there are generalizations for other target manifolds.

Suppose N is a compact, smooth Riemannian manifold of even dimension. It is called an almost-Kähler manifold if there exists a smooth tensor J on N that assigns to every point $y \in N$ an endomorphism $J(y)$ of the tangent space T_yN with the following properties:

- For every $y \in N$,

$$J^2(y) = -\mathrm{id}_{T_yN}.$$

- The tensor J is compatible with the metric on N in the sense that

$$\langle X, Y\rangle = \langle J(y)X, J(y)Y\rangle$$

  for every $y \in N$ and every pair of tangent vectors $X, Y \in T_yN$.
- It is parallel in the sense that

$$D^N J = 0,$$

  where D^N is the covariant derivative on N.

For the sphere S^2, the tensor given by

$$J(y)X = y \wedge X, \quad y \in \mathbb{S}^2,\ X \in T_y\mathbb{S}^2,$$

has these properties. Thus $\mathbb{S}^2$ is an example of an almost-Kähler manifold.

A natural generalization of the Landau-Lifshitz equation for maps into an almost-Kähler manifold is

$$\frac{\partial u}{\partial t} = a\tau(u) - bJ(u)\tau(u) \quad \text{in } (0,T) \times \Omega,$$

where

$$\tau(u) = \Delta u + A(u)(\nabla u, \nabla u)$$

is the tension field, as usual. Alternatively, we can write

$$a\,\frac{\partial u}{\partial t} + bJ(u)\frac{\partial u}{\partial t} = \tau(u) \quad \text{in } (0,T)\times\Omega.$$

If $a^2 + b^2 = 1$, it follows easily from the properties of J that these two equations are equivalent for any sufficiently regular u.

If $m \leq 4$, we can apply both the methods of Section 5.4 and Section 5.5 to a solution of such an equation. Combining them, we can prove a similar partial regularity result for solutions of this version of the Landau-Lifshitz equation as we obtained for the special case $N = \mathbb{S}^2$ in the previous section. We leave the details to the reader.

Bibliography

Adams, D. R. and Frazier, M. [1992]. *Composition operators on potential spaces*, Proc. Amer. Math. Soc. **114**, 155–165.

Aharoni, A. [2001]. *Introduction to the theory of ferromagnetism*, second ed., Oxford University Press, Oxford.

Almeida, L. [1995]. *The regularity problem for generalized harmonic maps into homogeneous spaces*, Calc. Var. Partial Differential Equations **3**, 193–242.

Bethuel, F. [1993]. *On the singular set of stationary harmonic maps*, Manuscripta Math. **78**, 417–443.

Bethuel, F., Brezis, H., and Hélein, F. [1993]. *Asymptotics for the minimization of a Ginzburg-Landau functional*, Calc. Var. Partial Differential Equations **1**, 123–148.

Bethuel, F., Brezis, H., and Hélein, F. [1994]. *Ginzburg-Landau vortices*, Birkhäuser, Boston.

Borchers, H.-J. and Garber, W.-D. [1980]. *Analyticity of solutions of the $O(n)$ non-linear σ-model*, Comm. Math. Phys. **71**, 299–309.

Chang, K.-C. [1989]. *Heat flow and boundary value problem for harmonic maps*, Ann. Inst. H. Poincaré Anal. Non Linéaire **6**, 363–395.

Chanillo, S. [1991]. *Sobolev inequalities involving divergence free maps*, Comm. Partial Differential Equations **16**, 1969–1994.

Chen, Y., Li, J., and Lin, F.-H. [1995]. *Partial regularity for weak heat flows into spheres*, Comm. Pure Appl. Math. **48**, 429–448.

Chen, Y. and Wang, C. [1996]. *Partial regularity for weak heat flows into Riemannian homogeneous spaces*, Comm. Partial Differential Equations **21**, 735–761.

Coifman, R. [1974]. *A real variable characterization of H^p*, Studia Math. **51**, 269–274.

Coifman, R., Lions, P. L., Meyer, Y., and Semmes, S. [1993]. *Compensated compactness and Hardy spaces*, J. Math. Pures Appl. **72**, 247–286.

Evans, L. C. [1991]. *Partial regularity for stationary harmonic maps into spheres*, Arch. Rational Mech. Anal. **116**, 101–113.

Fefferman, C. and Stein, E. M. [1972]. *H^p spaces of several variables*, Acta Math. **129**, 137–193.

Feldman, M. [1994]. *Partial regularity for harmonic maps of evolution into spheres*, Comm. Partial Differential Equations **19**, 761–790.

Friedman, A. [1964]. *Partial differential equations of parabolic type*, Prentice-Hall, Englewood Cliffs.

Gehring, F. W. [1973]. *The L^p-integrability of the partial derivatives of a quasiconformal mapping*, Acta Math. **130**, 265–277.

Giaquinta, M. [1983]. *Multiple integrals in the calculus of variations and nonlinear elliptic systems*, Annals of Mathematics Studies, vol. 105, Princeton Univ. Press, Princeton.

Giaquinta, M. and Modica, G. [1979]. *Regularity results for some classes of higher order nonlinear elliptic systems*, J. Reine Angew. Math. **311/312**, 145–169.

Gilbarg, D. and Trudinger, N. S. [1983]. *Elliptic partial differential equations of second order*, Second edition, Grundlehren der Mathematischen Wissenschaften, vol. 224, Springer-Verlag, Berlin.

Hélein, F. [1990]. *Régularité des applications faiblement harmoniques entre une surface et une sphère*, C. R. Acad. Sci. Paris Sér. I Math. **311**, 519–524.

Hélein, F. [1991a]. *Régularité des applications faiblement harmoniques entre une surface et une variété riemannienne*, C. R. Acad. Sci. Paris Sér. I Math. **312**, 591–596.

Hélein, F. [1991b]. *Regularity of weakly harmonic maps from a surface into a manifold with symmetries*, Manuscripta Math. **70**, 203–218.

Hubert, A. and Schäfer, R. [1998]. *Magnetic domains*, Springer, Berlin–Heidelberg–New York.

Iwaniec, T. [1992]. *p-harmonic tensors and quasiregular mappings*, Ann. of Math. (2) **136**, 589–624.

Iwaniec, T. and Martin, G. [1993]. *Quasiregular mappings in even dimensions*, Acta Math. **170**, 29–81.

Iwaniec, T. and Sbordone, C. [1994]. *Weak minima of variational integrals*, J. Reine Angew. Math. **454**, 143–161.

John, F. and Nirenberg, L. [1961]. *On functions of bounded mean oscillation*, Comm. Pure Appl. Math. **14**, 415–426.

Jost, J. [2002]. *Riemannian geometry and geometric analysis*, third ed., Universitext, Springer-Verlag, Berlin.

Ladyzhenskaya, O. A. and Ural'tseva, N. N. [1968]. *Linear and quasilinear elliptic equations*, Academic Press, New York.

Latter, R. H. [1978]. *A characterization of $H^p(\mathbf{R}^n)$ in terms of atoms*, Studia Math. **62**, 93–101.

Lions, J. L. and Magenes, E. [1972]. *Non-homogeneous boundary value problems and applications. Vol. I*, Springer-Verlag, New York.

Luckhaus, S. [1988]. *Partial Hölder continuity for minima of certain energies among maps into a Riemannian manifold*, Indiana Univ. Math. J. **37**, 349–367.

Morrey, C. B. [1966]. *Multiple integrals in the calculus of variations*, Grundlehren der mathematischen Wissenschaften, vol. 130, Springer-Verlag, New York.

Moser, J. [1961]. *A new technique for the construction of solutions of nonlinear differential equations*, Proc. Nat. Acad. Sci. U.S.A. **47**, 1824–1831.

Moser, R. [2003a]. *An epsilon-regularity result for generalized harmonic maps into spheres*, Electron. J. Diff. Eqns. **2003**, 1–7.

Moser, R. [2003b]. *Regularity for the approximated harmonic map equation and application to the heat flow for harmonic maps*, Math. Z. **243**, 263–289.

Moser, R. [2004] *Some remarks on the regularity of biharmonic maps in four dimensions*, preprint.

Nash, J. [1954]. *C^1 isometric imbeddings*, Ann. of Math. (2) **60**, 383–396.

Nash, J. [1956]. *The imbedding problem for Riemannian manifolds*, Ann. of Math. (2) **63** (1956), 20–63.

Nirenberg, L. [1959]. *On elliptic partial differential equations*, Ann. Scuola Norm. Sup. Pisa (3) **13**, 115–162.

Price, P. [1983]. *A monotonicity formula for Yang-Mills fields*, Manuscripta Math. **43**, 131–166.

Rivière, T. [1995]. *Everywhere discontinuous harmonic maps into spheres*, Acta Math. **175**, 197–226.

Schoen, R. and Uhlenbeck, K. [1982]. *A regularity theory for harmonic maps*, J. Differential Geom. **17**, 307–335.

Schoen, R. and Uhlenbeck, K. [1983]. *Boundary regularity and the Dirichlet problem for harmonic maps*, J. Differential Geom. **18**, 253–268.

Simon, L. [1996]. *Theorems on regularity and singularity of energy minimizing maps*, Lectures in Math. ETH Zürich, Birkhäuser, Basel.

Stein, E. M. [1970]. *Singular integrals and differentiability properties of functions*, Princeton University Press, Princeton.

Stein, E. M. [1993]. *Harmonic analysis: real-variable methods, orthogonality, and oscillatory integrals*, Princeton University Press, Princeton.

Struwe, M. [1985]. *On the evolution of harmonic mappings of Riemannian surfaces*, Comment. Math. Helv. **60**, 558–581.

Struwe, M. [1988]. *On the evolution of harmonic maps in higher dimensions*, J. Differential Geom. **28**, 485–502.

Strzelecki, P. [2003]. *Regularity of generalized shpere valued p-harmonic maps with small mean oscillations*, Manuscripta Math. **112**, 473–487.

Index